William Henry Bennett

Clinical Lectures on Abdominal Hernia Chiefly in Relation to

Treatment Including the Radical Cure

William Henry Bennett

Clinical Lectures on Abdominal Hernia Chiefly in Relation to Treatment Including the Radical Cure

ISBN/EAN: 9783337427863

Printed in Europe, USA, Canada, Australia, Japan

Cover: Foto ©berggeist007 / pixelio.de

More available books at **www.hansebooks.com**

ON

ABDOMINAL HERNIA

CHIEFLY IN RELATION TO TREATMENT

INCLUDING

THE RADICAL CURE

WILLIAM H. BENNETT, F.R.C.S.

SURGEON TO ST GEORGE'S HOSPITAL; MEMBER OF THE BOARD OF EXAMINERS
FOR THE FELLOWSHIP ROYAL COLLEGE OF SURGEONS OF ENGLAND
AND LECTURER ON CLINICAL SURGERY IN ST GEORGE'S
HOSPITAL MEDICAL SCHOOL

WITH TWELVE DIAGRAMS

LONDON

LONGMANS, GREEN, AND CO.

AND NEW YORK: 15 EAST 16th STREET

1893

ABDOMINAL HERNIA

CLINICAL ... DIAGNOSIS AND TREATMENT

including

THE RADICAL CURE

WILLIAM H. BENNETT, F.R.C.S.

LONDON

J. & A. CHURCHILL

PREFACE

THESE LECTURES, which treat principally of hernia in the groin, were delivered at St. George's Hospital at irregular intervals, the subject of each lecture being from time to time suggested by the occurrence of cases under my care in the wards.

They are, therefore, in no sense systematic or exhaustive, and deal only with questions upon which, after much practical experience and careful observation, I feel justified in expressing definite views.

The greater part of the purely elementary matter necessarily included in lectures addressed to students has been omitted; some, however, is here and there retained, mainly for the sake of clearness.

With this exception, and a few unimportant changes in the earlier lectures, in order to bring them up to date, they are published as originally given.

The Appendices to Lectures III., VI., and XI.

have been added during the preparation of the volume for the press.

I do not suppose that anything can be told of an affection so common as hernia which will be new to persons whose knowledge of the subject is fully matured. I have nevertheless ventured upon the publication of the lectures in their present form because, when somewhat younger in my career and less experienced than now, I found it difficult, and in truth sometimes impossible, to obtain from authentic sources precise information upon some of the points which I have endeavoured to discuss.

W. H. B.

1 CHESTERFIELD STREET, MAYFAIR :
March 1893.

CONTENTS

— ⁘ —

The Lectures are arranged with reference to the subjects treated of, and not in the order in which they were delivered.

LIST OF CASES FULLY DESCRIBED IN THE TEXT
AND USED AS CLINICAL EXAMPLES TO ILLUS-
TRATE SOME OF THE POINTS DISCUSSED

LIST OF DIAGRAMS

CLINICAL LECTURES

ON

ABDOMINAL HERNIA

———◆◇◆———

LECTURE I

ON THE DECEPTIVE NATURE OF THE SYMPTOMS OF STRANGULATION

(Published in the *Lancet*, December 19, 1891)

SYNOPSIS—Introductory remarks—STRANGULATION WITHOUT SYMPTOMS OF INTESTINAL OBSTRUCTION—Clinical example (No. I.), strangulated omental hernia—Clinical example (No. II.), strangulated vermiform appendix—Remarks on these cases—Absence of true hernial impulse—Characteristics of the same—Difference of impulse in herniæ containing bowel and those composed of omentum—Reason for this—Relation of absence of impulse to treatment—Putrefactive crackling in sac—Treatment of strangulated vermiform appendix—Probable explanation of absence of intestinal obstruction in cases described, and other cases in which it may occur—SIGNS OF STRANGULATION OCCURRING WITHOUT HERNIA—Clinical example (No. III.), glandular abscess in crural canal, simulating strangulated hernia—Remarks on this case—Existence of intestinal obstruction not mere coincidence—Possible explanation of case—Symptoms of deceptive nature not confined to hernia.

THE symptoms of strangulated hernia are generally so clear and uncomplicated that a mistake in the diagnosis of the condition can hardly occur, excepting from ignorance or unpardonable carelessness.

B

At the same time, cases are occasionally seen, the symptoms of which differ so much from the classic descriptions of the books that they deserve careful study. Three such cases have been recently under my care in the hospital, and I therefore propose to devote the present lecture to their consideration.

In two instances the symptoms of intestinal obstruction were entirely wanting; and in the third, the most interesting of the series, every symptom of strangulation of a femoral hernia developed, and yet, upon operating, no rupture was found, nor had one apparently ever been present.

CASE I.—*Strangulated omental hernia, in which there was no interference with the action of the bowels ; herniotomy ; commencing gangrene of contents of sac ; radical cure ; recovery.*

W. L——, a labourer, thirty-two years of age, was admitted into the Belgrave Ward on October 22, 1890. He had been ruptured since he was fifteen years old; for five years after the original discovery of the hernia he had regularly worn a truss, but for ten years he had used no support of any kind. The rupture had always been reducible till four days before his admission, when he suddenly felt some pain in the scrotum, found the rupture down, and could not put it back. The hernial swelling gradually increased in size, and the pain became more marked. There was neither vomiting nor nausea, and the bowels acted with daily regularity up to and including the morning of his admission. He came to the hospital on account of the pain, and for no other reason. He had walked straight from his work to seek relief, and great pressure was required to prevent his returning to it. On admission the patient complained of pain about the scrotum, with a feeling of dragging and weight around the umbilicus. There was a scrotal tumour the size of a small cocoanut, rather tense, a little tender, and irreducible. When I saw him, two hours after admission, almost all the pain had disappeared, and he begged to be allowed to leave the hospital at once. On examination I found the tumour of the size above mentioned, and in much the same condition. When the patient coughed the tumour moved freely under the hand, but the true hernial impulse was wanting. There was no constitutional disturbance, the temperature was only 99°, the pulse 76. Beyond the absence of the hernial impulse there was nothing distinctive of strangulation of the hernia. Herniotomy was at once performed. On opening the sac a quantity of fluid, at first clear and then blood-stained, escaped. The sac contained also a large mass of omentum congested throughout, at the back of which was a small area of commencing gangrene. There was a tight stricture at the internal ring, which required free division. The mass of omentum was ligatured and removed, with the exception of a pad of the healthiest part, which, after careful cleansing with a solution of 1 in 1,000 of corrosive sublimate, was stitched across the ring in the manner described by me in 'The Lancet' of Sept. 12, 1891 (See Lecture XI.). The sac was subsequently invaginated, and the pillars of the ring approximated by silk suture in the usual manner. The progress towards recovery was uninterrupted, and the patient left the hospital well on November 26.

CASE II.—*Painful tumour over the saphenous opening; diarrhœa; exploratory operation; strangulated hernia of vermiform appendix, which was gangrenous: removal of appendix; recovery.*

Sarah B——, aged 53, was admitted into the Princess Ward on November 24, 1890, suffering from what was supposed to be inflammation of some glands in Scarpa's triangle. She had always enjoyed good health until a year previously, when, in consequence of ' a strain,' a swelling came below the right groin. This swelling had ever since been troublesome and uncomfortable, becoming painful when she walked much ; it was also tender, especially towards the end of the day. Four days before coming to the hospital she felt rather suddenly ' out of sorts,' and vomited. The swelling soon afterwards became painful and increased a little in size. Diarrhœa immediately followed, and continued up to the time of her admission. I saw the patient directly after she had been sent to bed. She was obviously very ill. The expression was anxious, the tongue whitish and rather dry, the temperature 102°, and the pulse quick. In the right groin, a little above the saphenous opening, was a hard swelling, rather irregular in shape, and presenting the exact appearance of a mass of inflamed glands. It was irreducible and tender ; there was no impulse, and the skin was not infiltrated or unhealthy. On careful manipulation, there could be felt in the tumour a very faint crackling sensation. I therefore concluded that, in spite of the diarrhœa, the tumour was a hernia with gangrenous contents (probably omental), and possibly the condition known as Richter's hernia. The mass was therefore at once explored. After cutting through a quantity of inflammatory material, a small peritoneal sac was laid open, from which came a little fetid fluid. At the bottom of this sac lay a black wormlike body, which formed its entire contents, and proved to be the vermiform appendix, gangrenous and very tightly strangulated at the femoral ring. The parts having been thoroughly cleansed with a solution of corrosive sublimate (1 in 1,000), the stricture was freely divided, the gangrenous part of the appendix being cut off and got out of the way. The incision made in dividing the stricture was then extended into the abdomen sufficiently to allow the cæcum to be drawn into the wound. The stump of the appendix was then removed through perfectly healthy tissues about half an inch from the cæcum. The peritoneal surfaces were inverted without difficulty, and retained in apposition by means of three Lembert's sutures. The parts, having been again thoroughly cleansed, were then returned into the abdomen, and the wound in the parietes brought together with silkworm gut sutures. Rapid recovery followed. The diarrhœa ceased, and the temperature became normal within forty-eight hours of the operation. She was discharged well on December 22.

It will be noticed that in one of these cases constipation and vomiting, two of the symptoms most insisted upon as affording certain evidence of strangulation, were absent; and in the other not only was there no intestinal obstruction, but, on the contrary, diarrhœa was present. Yet, in each a portion of the abdominal contents was so tightly constricted as to produce a condition of gangrene, partial in one instance and complete in the other.

Now, the final diagnosis in each of these cases depended upon a single symptom the recognition of which was fairly simple after a careful examination of the local conditions.

In both instances an irreducible painful and tender tumour existed, in which some increase of size and tension had been noticed. These symptoms, although suggestive of strangulation, were, I need hardly say, not in the least conclusive, as a hernia may at any time temporarily increase in size, and become painful, without being strangulated.

The conclusive symptom in Case I. was the *entire absence of the real hernial impulse*—a fact quite sufficient, when associated with the suggestive symptoms just mentioned, to make the diagnosis certain, in spite of the singular absence of constipation or vomiting. When I first went to see the patient I was told that marked impulse could be felt, but upon examination it was clear to me that, although the tumour moved under the hand rather freely when the patient coughed, there was no true hernial impulse in the proper sense of the word. I therefore at once operated, with the result seen in the description of the case.

The detection of hernial impulse in a case like this is obviously a matter of the first importance, because, if this impulse be present the treatment by some temporising method, such as the local application of cold by the ice-bag or otherwise, may not be in itself improper. If, on the other hand, the impulse be absent, as happened here, such a treatment is not only useless but positively dangerous, inasmuch as its only effect can be to postpone the inevitable herniotomy until the patient's condition has become less favourable for operation.

At the risk, therefore, of the matter being perhaps thought too trivial and elementary to be dealt with in this place, I venture to discuss for a moment the symptom which is called 'impulse on coughing,' because I am not sure that the meaning of the term is always understood, and also because I have more than once seen the ice-bag applied and violent taxis used in cases of hernia on the strength of this impulse being present when it certainly did not exist.

The impulse in ordinary non-strangulated hernia, whether the contents of the sac consist of omentum or bowel, is *expansile* in character—that is to say, the tumour when the patient coughs or strains not only rises under the hand, but expands in size. In hernial tumours containing bowel this sudden increase in the bulk is principally due to the additional quantity of gas, &c., which is driven into the herniated portion of gut by the act of coughing or straining. In omental hernia the expansion is partly due to the sudden turgescence in the omental vessels, and partly to the increase of tension in the sac caused by the cough or strain. Naturally, therefore, the amount of the expansion is relatively greater in herniæ containing bowel than in those composed only of omentum. So much is this the case that in a certain number of instances an experienced practitioner may form a fairly accurate idea of the nature of the contents of a hernial sac by the careful estimation of the amount of expansion in the impulse. The only cases of ordinary non-strangulated hernia in which this expansile impulse is absent, except upon very careful examination indeed, are those in which the hernia consists only of very old indurated omental masses, which completely fill the sac and block the opening between it and the abdominal cavity.

In these cases, if the examination be confined to the lower portion of the tumour, the expansion may not be perceptible; but any more recent portions of the mass lying near the ring will generally afford the expansile symptom to a person possessing anything like a delicate sense of touch who examines the hernia in the right manner—*i.e.* by grasping the lower part of the tumour in one hand whilst its neck is held between the fingers and thumb of the opposite hand, the patient being at the same time made to cough or strain.

In strangulated hernia it is important to understand that absence of impulse does not necessarily mean, as some seem to think it does, *immobility* during coughing, for a hernia, even if tightly strangulated, will often move freely under the hand, especially if it be omental. This movement, however, is rather of the nature of a jump or jerk, and is never expansile. There is no symptom which has a more practical bearing upon the treatment of strangulated hernia than the expansile character of this impulse.

It may be safely held as a surgical dictum, that *every case of hernia in which any change has taken place in the condition of the tumour, such, for example, as increase of size or tension, whilst expansile impulse is absent, should be regarded as strangulated*, and treated accordingly, without waiting for the onset of vomiting or other sign of intestinal obstruction. Had I been less certain of my diagnosis in this case, I might possibly have been tempted to wait for further symptoms, the first of which would have been vomiting, produced, in all probability, not by intestinal obstruction, but by peritonitis. A respected teacher of mine used frequently to say, ' Never let a patient with a strangulated hernia vomit twice.' Personally, I would go further than this, and ask why a patient with a strangulated hernia should, *as a matter of course*, be allowed to vomit at all.

Although vomiting is nearly always present in the cases with which we have to deal in hospital practice, it is certainly not a necessary condition in cases of strangulation (especially when gut is not involved) in the earliest stage, which may occasionally be recognised by the means I have indicated before the onset of this symptom. When once the diagnosis of a strangulated hernia has been made, herniotomy cannot be too soon performed, and the use of the ice-bag (or other temporising methods of treatment) should be regarded as a cause of harmful delay, as the surgeon ought always to have in view not only the relief of the constriction, which is, of course, his first care, but also the possibility of helping the patient further by the performance of an operation for the radical cure of the rupture.

Any unnecessary delay in operating caused by the waiting for the development of further symptoms—such, for instance, as vomiting—may so reduce the powers of the patient as to render only the shortest of operations feasible, and perhaps imperil the ultimate chance of recovery ; and short of such a serious result as this, it may allow the hernia to be so much damaged by the actual constriction, or lead to such an extension upwards of septic changes involving the peritoneum, that closure of the sac or canal may be indiscreet or entirely unjustifiable.

In Case II. the distinctive symptom was not the absence of impulse, but the existence of putrefactive crackling in the tumour, which could only have proceeded under the circumstances from gangrenous tissues. The probability of those tissues being some of the contents of the abdominal cavity was so great as to amount almost to a certainty. The difficulty about this part of the matter lay in the extreme indistinctness of the crackling ; for although I could feel it without doubt myself, I was unable actually to demonstrate the sensation to all of those present.

The tumour itself presented the irregular shape often seen in inflamed glandular masses. There was tenderness with some heat and stabbing pain, but there was no redness. These symptoms, together with the fever and diarrhœa not infrequently met with in septic conditions, might easily have led to a diagnosis of commencing suppuration in the femoral glands, if the emphysematous crackling had escaped notice. Such a mistake might have led to serious results had any delay in the exploration of the tumour followed in consequence. The treatment adopted after the exposure of the gangrenous tissues—viz. the extension of the operation wound (making, in fact, a small abdominal section) and the removal of the appendix through healthy parts—was clearly right.

The case is a rare one. In the 'Lancet' for

1889, vol. i. p. 627, Mr. Annandale records an instance in which a strangulated and perforated appendix cæci occupied a small hernial sac. He removed the strangulated portion of the appendix and fixed its base, which was very adherent, by means of sutures between the pillars of the ring. In that case the local conditions which existed may have made the course adopted necessary, but when possible it is undoubtedly better to remove the appendix and return the carefully cleansed parts into the abdominal cavity, in order to avoid anchoring the cæcum to the parietes in a way which may at any time lead to trouble.

In connection with Case I. the complete absence of constipation or other intestinal complication is remarkable, for it is a notorious clinical fact that strangulation of herniated omentum is generally associated with constipation of the most obstinate kind, and frequently with all the other symptoms of intestinal obstruction. The absence of these symptoms in this case is at first sight all the more curious when the severity of the strangulation is considered. The explanation of the matter is, however, I fancy, not difficult if due regard be paid to the fact that the acuteness of the symptoms in these omental cases bears a direct relation to the recent nature of the hernia—*i.e.* the more recent the hernia, the more acute are the symptoms of intestinal obstruction. The strangulation in old cases of omental hernia is almost invariably connected with the descent into the sac of a new piece of omentum, so that, although the hernia itself may be an old one, the actual strangulation is caused by the recently herniated part, and the symptoms are proportionately acute. In the case under discussion there was no recent omentum in the sac, and the constriction of the old indurated omentum did not, therefore, appear to have caused irritation enough to react upon the bowel and modify its function.[1]

The cause of the strangulation, so far as it was possible to judge from the condition of the parts removed, appears to have had its origin in extensive thrombosis of the large veins (? from injury) in the lower part of the omental mass, followed by swelling and rapid effusion into the sac. In this way the relation of the neck of the hernia to the margins of the ring might easily have been altered sufficiently to produce constriction and ultimately complete strangulation.

The diarrhœa in the second case, although misleading from a diagnostic point of view, was not altogether inconsistent with the condition of parts found at the operation. In Mr. Annandale's case, to which I have referred, the signs of intestinal obstruction were absent, but there was no diarrhœa;

[1] Further reference to this point will be found in Lecture VII.

this symptom may, however, be met with in the strangulation in a hernia of an intestinal diverticulum (Littré's hernia), and may also occur in partial enterocele (Richter's hernia) when the portion of the circumference of the gut which is strangulated is very small.

It is here interesting to note that strangulation of the appendix cæci in a hernia will sometimes give rise to very acute symptoms of intestinal obstruction, although it may be neither perforated nor gangrenous; in· the 'Lancet' for 1880 (vol. i. p. 801) Mr. Pick, for example, gives an account of such a case.

The third case to which I wish to direct attention is in some respects unique in interest, and it is not a little remarkable that the patient should have been admitted and operated upon within a few hours of the occurrence of such an unusual case as the second of the two just considered.

CASE III.—*Symptoms of strangulated femoral hernia, with acute intestinal obstruction, caused by an abscess in the crural canal; operation; immediate and complete relief.*

A. P——, a woman, 36 years old, was admitted into Princess Ward on November 24, 1890. She had always been in good health in every respect until six months before her admission. About that time she noticed a 'lump' in the right groin, which varied in size occasionally, but never entirely disappeared. Sometimes, after a hard day's work, the swelling increased very considerably. There was usually a feeling of discomfort, and often there was slight tenderness. A medical man, who had once been called in, attempted to reduce the tumour, and recommended that a truss should be used, which was, however, not done. On November 21 (three days before admission), the swelling, without any ostensible cause, became rapidly larger than it had ever before been. Pain and tenderness followed; vomiting commenced almost immediately, and recurred frequently up to the time of her arrival at the hospital. The bowels acted for the last time on the morning of the 21st. On admission the patient appeared to be very ill; the pulse was quick, the tongue dry, and the temperature 102·4°. There was copious vomiting of ill-smelling semistercoraceous material. In the right groin was a rounded mass as large as a full-sized walnut, tense, painful, and so tender that any accurate examination was impossible. The pain was apparently characteristic of a strangulated hernia, for it extended over the lower part of the abdomen, and, passing up from the right groin, it constantly dragged upon the umbilicus. A strangulated femoral hernia was therefore diagnosed, and what was thought to be a herniotomy proceeded with at once. After cutting down through some inflamed tissues, which bled freely, a rounded fluctuating swelling was exposed, having none of the appearances seen about a hernial sac. From this there came, after incision, about two teaspoonfuls of thinnish pus, with a small cretaceous mass, evidently from a caseating gland. The lining membrane of this abscess was easily scraped away, leaving a distinct capsule with clean walls. No hernia of any kind was found, nor had the abscess (which was dis-

tinctly made out to occupy the crural canal) any connection whatever with the peritoneal cavity. The plug
of connective tissue (septum crurale) in the femoral
ring was intact, undisturbed, and natural in all respects, proving conclusively that no hernia could have
at any time existed at this spot. The wound was
stitched up and dressed antiseptically as usual. Complete relief followed the operation. There was no subsequent vomiting; the pain ceased, and the temperature
fell to normal. The bowels acted spontaneously in
about forty-eight hours, and rapid recovery followed
without a hitch of any kind.

It would be difficult to produce a case showing
more admirably than this one does the impossibility
of being *always* right in assuming that any sequence
of symptoms, however perfect, is absolutely distinctive of such a condition as a strangulated hernia,
which is usually held to be easy of diagnosis. It is
well that cases of this kind should occasionally be
met with, since they serve as a wholesome warning
against taking anything for granted, even in symptoms which seem perfectly clear, and show the
necessity of carefully weighing every case entirely
upon its own merits. It is true that in the present
instance the diagnosis of strangulated hernia when a
hernia did not exist brought no harm to the patient,
but rather the reverse, since it led to speedy exploration of the tumour and resulted in the immediate
relief of urgent symptoms. In a different class of
case it might, however, be quite otherwise, for the

awkward fact remains that a train of symptoms precise in character and detail, and identical in all respects with those universally taught to be diagnostic of a certain pathological state, existed, whilst the disease itself was altogether absent. I must frankly admit that I am at a loss for any reasonable explanation of the exact relation between the local conditions found in this case and the symptoms of intestinal obstruction.

Every surgeon of experience occasionally meets with a case in which the signs of strangulated hernia are produced, in the absence of any such condition, by the irritation of structures connected with the peritoneum (*e.g.* the inflammation of an empty hernial sac or the irritation or inflammation of an undescended testis); but I have never myself seen nor have I heard of any condition, inflammatory or otherwise, *having no immediate connection with the peritoneum,* which has given rise to a sequence of symptoms so perfect as was met with here. The immediate relief which the operation afforded fairly negatives the possibility of the coexistence of the abscess and the intestinal symptoms being merely accidental. It appears, therefore, so far as I understand the matter at present,[1] that there is nothing to be found to answer better as a cause for the relation between the symptoms than reflex nervous irritation,

---

[1] See additional remarks on this case in Appendix to Lecture III.

that fertile resource upon which we are apt to rely a great deal too much for the explanation of doubtful causes. In this instance, at all events, I must say that the theory seems too far-fetched to be of much practical account.

In conclusion, let it be remembered that symptoms of a deceptive nature are not confined to the class of ailment to which the cases discussed here belong, but occur in connexion with many other pathological states. Such symptoms are less likely to seriously mislead in rare or out-of-the-way cases than when they are met with unexpectedly, as they often are in diseases which are being continually dealt with, and which may consequently come to be regarded as universally ordinary and straightforward.

It would, I think, be impossible to find a series of cases to better illustrate this point than that which forms the basis of the present lecture.

LECTURE II

ON THE DECEPTIVE NATURE OF THE SYMPTOMS OF STRANGULATION (continued)

(Not previously published)

SYNOPSIS—EXISTENCE OF STRANGULATION NOT ALWAYS INCOMPATIBLE WITH EXPANSILE IMPULSE IN HERNIA—Clinical example (No. IV.), strangulated internal hernia with impulse in scrotal sac—Clinical example (No. V.), hernia strangulated in the sac, free impulse—Remarks on these cases—Clinical importance of the fact that free impulse may exist in the hernial tumour in some cases of strangulation—Case IV. probably unique and not to be explained upon ordinary grounds—Cause of the impulse in scrotal sac—Case V. not rare, but important—Reason of *absence* of impulse in ordinary cases of strangulation—cause of its *presence* in strangulation in the sac—Diagnosis rarely difficult—Indications as to diagnosis and treatment—STRANGULATION IN THE SAC NOT ASSOCIATED WITH IMPULSE IN TUMOUR IF THE HERNIA IS ALSO STRANGULATED AT ABDOMINAL RING—Clinical example (No. VI.), strangulation in the sac by a Meckel's diverticulum and also at the ring—Explanation of the case—TREATMENT OF STRANGULATED HERNIA COMPOSED OF LARGE MASSES OF MATTED AND ADHERENT INTESTINE—Justification for author's treatment in this case—Importance of returning gut into abdomen, however adherent—Reference to case showing danger of leaving gut in the sac after herniotomy.

IN my last lecture, in calling attention to the deceptive nature of some of the recognised symptoms of strangulated hernia, I especially dwelt upon the importance of the *absence of the true hernial impulse as a perfectly certain sign of strangulation, although vomiting and other signs of intestinal obstruction may be wanting.* I now propose to describe two cases which illustrate exactly the opposite condition, as they show that *strangulated gut or omentum may exist in a hernial tumour although the true expansile impulse is present and well marked at the same time.*

Arising out of these cases are some other points of considerable import, not only from their intrinsic interest but also because they have a direct bearing upon treatment.

CASE IV.— *Symptoms of intestinal obstruction; scrotal tumour with free impulse on coughing; herniotomy; contents of sac fluid only; abdominal section; rare form of internal hernia; recovery.*

I. S., aged 60, was admitted into the Belgrave Ward under my care on June 10, 1890, and gave the following history. He had been ruptured, so far as he knew, for about seven years, the hernia always disappearing when he lay down. No truss had been worn.

Three days before coming to the hospital the tumour seemed, after a sudden ' stitch ' in the left side, to grow larger gradually, and whilst coughing (he was the subject of chronic bronchitis), became painful. The same night he noticed for the first time that the hernia did not lessen in size on his lying down when he went to bed. The pain soon became worse, in the scrotum especially, and, passing up into the belly, seemed to concentrate itself just to the left of the umbilicus. Vomiting soon followed and frequently recurred, causing irresistible desire to strain for the purpose of passing a motion, but without effect, a little flatus only on one occasion having been forced out.

A medical man who was called in attempted to reduce the ' rupture ' but failed.

On admission the patient was collapsed ; the pulse small and quick; the skin clammy; the aspect bluish and distressed.

There was profuse vomit of dark-greenish, ill-smelling material.

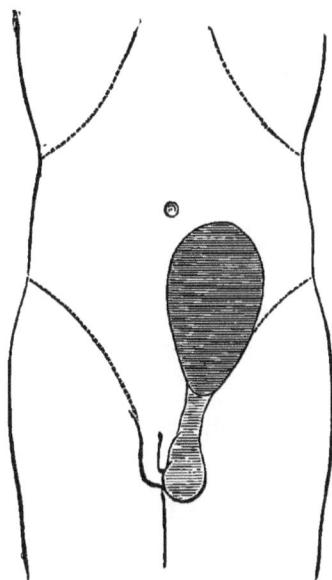

FIG. 1.—THE SHADED AREA INDICATES THE EXTENT OF DULNESS.

The urine (specific gravity 1020) contained one-fourth its bulk of albumen.

The left inguinal region was altogether somewhat full and very tender to the touch. The fulness extended down into the scrotum as an irreducible, not very tense, swelling, feeling like a rather flaccid hernial sac. In this there was distinct, easily demonstrated, impulse on coughing, which was expansile in character.

Extending upwards from the groin towards the

region of the umbilicus was an area of dulness, the extent of which is seen in the accompanying diagram. (Fig. 1.)

Under ether the scrotal tumour was laid open and explored as in an ordinary herniotomy. A peritoneal sac was exposed which contained about two ounces of fluid, very slightly blood-stained. The sac was otherwise empty, and appeared at first to be entirely cut off from the abdominal cavity, as no opening could be detected by passing the finger to its uppermost limit.

It was, however, noticed that the inguinal fulness and abdominal dulness remained unchanged, and now upon again examining the upper end of the scrotal sac I found at the back part a distinct fold, behind which I was enabled after a little manipulation to pass my finger; a gush of fluid deeply blood-stained immediately occurred, and I should say that certainly not less than a pint escaped, which was of quite a different character from that which the scrotal sac contained.

The finger end passing beneath the membranous fold, entered what seemed to be a large smoothly-lined cyst, and on firm pressure backward being made on the umbilical region the finger, pushed upwards in this intra-abdominal cyst, impinged upon a rounded tense mass which was evidently a constricted knuckle of gut.

Abdominal section was then performed by prolonging the scrotal wound upwards, and the following condition of parts was found :—

Projecting into. a large peritoneal intra-abdominal sac, through a circular opening about the size of a shilling-piece, was a knuckle of deeply-congested, almost black intestine about six inches long.

The sac was of sufficient size to hold fully a pint, and communicated by a constricted neck with the scrotal sac ; at about the level of the internal ring there crossed the channel of communication between the two sacs, a fold, or rather flap (under which my finger had passed), which seemed of just such a size and shape as to be capable of closing the canal in the manner of a valve, the concave side of which presented towards the belly. (Fig. 2.)

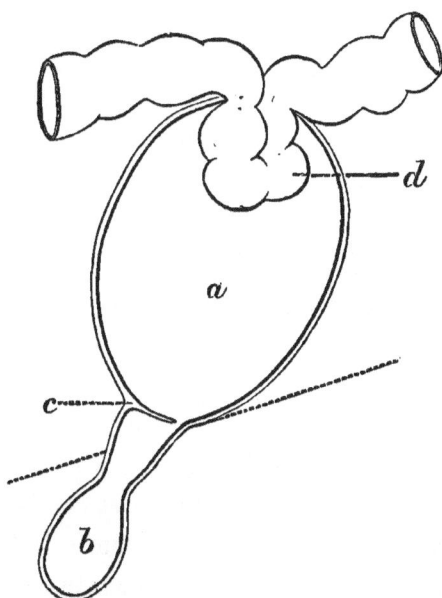

FIG. 2.—SHOWING GENERAL ARRANGEMENTS OF PARTS IN THIS CASE.

a, Abdominal sac with knuckle of gut *d* strangulated at upper part.
b, Scrotal sac.
c, Valve-like flap at junction of upper and lower sac, which, closed by pressure of fluid from above, completely shut off scrotal sac.
The dotted line indicates the level of Poupart's ligament.

The strangulated bowel was released by the division of the constricting ring, and the communication between the abdominal and scrotal sacs having been obliterated by suture and ligature, the operation wound was united by suture. Vomiting occurred (presumably from the effects of the ether) twice during the first twenty-four hours after the operation, and considerable trouble arose from the bronchitis; otherwise the progress towards recovery was uninterrupted. The bowels acted spontaneously on the sixth day after the operation.

All the stitches had been removed by June 21, and the patient left the hospital perfectly well on July 30.

CASE V.—*Intestinal obstruction of eleven days' duration; scrotal hernia with free impulse on coughing; herniotomy; strangulated gut in lower end of tri-locular sac; death from collapse.*

R. M——, a man aged 51, was admitted into Harris Ward on September 26, 1891, with the following history. To the best of his knowledge, he had been ruptured only four years, but he 'always knew that the ' right testicle was larger than the left.' No truss had been worn, and the rupture was always down. Eleven days before being brought to the hospital, he was seized suddenly with acute pain in the pit of the stomach whilst straining at stool. Vomiting followed almost directly. The pain 'hung about' him for three or four days, being sometimes in the belly and some times in the 'testicle,' which he considered had at the same time swollen. The vomiting occurred frequently, but only after taking food, until a week before his admission, when the pain became so severe that he took to his bed and treated himself for biliousness. The vomiting then became more frequent, and occurred independently of the taking of food.

For four days the vomiting had been very frequent, and the material stinking. The pain became less, but the abdomen gradually grew very large and extremely tender over the lower part. No action of the bowels had occurred since the outset of the attack.

On admission.—A flabby-looking subject, greatly collapsed. Skin cold and clammy. Face congested. Breath stinking. Tongue dry and hard. There was frequent vomiting of stercoraceous material; the abdomen was much distended, and there was tenderness over the hypogastrium.

There was a right scrotal hernia of moderate size and irreducible, in which there was free expansile impulse ; at the lower part of the tumour was a hard mass about the size of a hen's egg, closely connected with the testicle, which seemed normal. The pain which when I saw him was entirely abdominal, the tenderness which was also confined to the belly, the whole hernia being so insensitive that free attempts at reduction had given rise to no discomfort, and the extreme severity of the symptoms together with the manner of their onset, certainly to some extent pointed to the obstruction being in the abdomen rather than in the hernia, so much so that the house surgeon had prepared everything for abdominal section. It was clear, however, that the symptoms were not improbably connected with the strangulation of the gut in the lower end of the sac. At all events, the obvious course was to explore the hernial tumour.

This I accordingly did, and exposed in the sac some very slightly distended gut, which was perfectly healthy excepting towards its lower end, where a recent adhesion glued it to the sac. Below this the gut passed through a small round opening, not larger than a sixpenny-piece, with a very sharp margin which tightly gripped the bowel, which, after division of the constricting structure, was found to occupy together with a little dark fluid, a rounded pouch at the lower end of the sac which was quite unconnected with the tunica vaginalis. The surface of the gut was indented with a line of ulceration produced by the sharp edge of the constriction, and the knuckle itself, although not actually gangrenous, was on the point of becoming so.

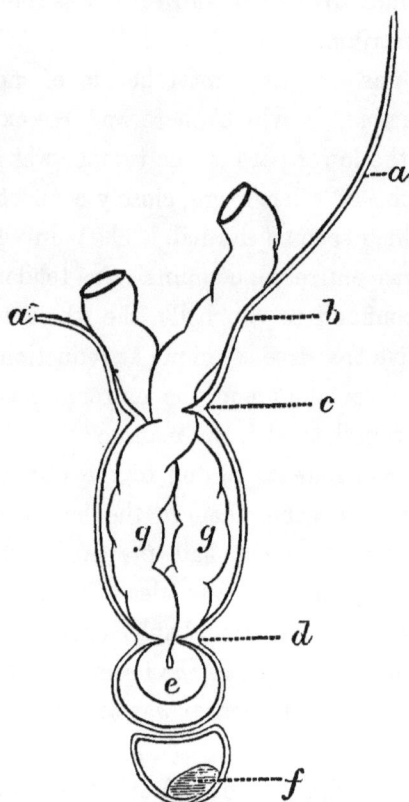

Fig. 3.—Diagram of the condition of parts found in Case V., showing Strangulation in the Sac.

a, a, General peritoneum passing down to form sac.
b, Level of internal ring.
c, Upper diaphragm in sac which caused no injurious pressure on bowel.
d, Lower diaphragm strangulating loop of gut marked *e*.
f, Testicle in tunica vaginalis which, quite distinct from hernial sac, contained some fluid.
g, Unstrangulated bowel in upper part of sac, in which the true hernial impulse could be felt, and which entirely concealed the small constricted knuckle below.

After the relief of the strangulation, the reduction of the hernia could not be effected until an annular membranous diaphragm, a little below the level of the internal ring, which seemed in no way to press upon the bowel, had been divided, after which the reduction was perfectly easy.

The wound was then closed by suture, a large drainage-tube having been placed in the abdomen close to the returned bowel. Although the man did not actually rally after the operation, no further vomiting occurred. He sank exhausted eighteen hours later. The condition of parts found at the operation are shown in fig. 3.

At the *post-mortem* examination the operation wound was healthy and nearly healed ; the edges of the incision in the hernial sac were firmly united with lymph.

'Lying close to the right internal ring, but not ' adherent to it, is a piece of ileum, $4\frac{3}{4}$ inches long, ' blackish and gangrenous (? decomposed), and covered ' with a few flakes of lymph. The mesentery has also ' been strangulated, but there is no gangrene.' The damaged gut 'commences at 16 inches from the cæcum, ' and at a distance of 10 inches from the cæcum the ileum ' presents a black pigmented ring due to former strangu- ' lation.'

From the description of these cases it will be seen that in both there was undoubted intestinal obstruction, whilst there existed in each case distinct expansile impulse in the hernial sac.

Taken in connection with the lesson taught by the cases I. and II., described in Lecture I., this fact is of much clinical importance; for, as I have already said, it demonstrates very clearly that, although the *absence* of the true hernial impulse when associated with any local change in the rupture is pathogno-

monic of strangulation, the *presence* of this same
impulse in the hernia, when symptoms of intestinal
obstruction exist, affords *in itself* no grounds for
concluding that the rupture is not strangulated.

It is impossible to insist too strongly on this
clinical truth, for I have in my own practice met
with cases in which the vomiting and abdominal
pain from the strangulation of a hernia *in the sac*
have been treated as symptoms of ' acute biliousness '
*on the strength of the presence in the hernia of the
characteristic impulse in coughing,* which was held to
negative the existence of strangulation, although
unmistakeable changes had occurred in the hernial
tumour.

Case IV. is of such a rare kind that I have
never seen anything at all like it, nor can I find a
record of any similar case. The case was not one
of *réduction en bloc,* and, from the appearance of
the parts it was, I think, pretty evident that this
complicated sac owed its existence to congenital
causes. The scrotal sac was the unclosed tunica
vaginalis, and identical, therefore, with the sac of
a congenital hernia. By what morphological, or,
indeed, pathological process, the abdominal sac was
formed, I am unable to say, for it was, so far as
could be made out, not merely a greatly distended
and thinned upper end of a curious inguinal sac, but
a distinct development.

A very interesting point, also, which I shall discuss in another lecture, is the valve-like flap at the junction of the upper and lower portions of the sac, which, under pressure, completely shut off the two portions from one another in such a manner that no fluid could pass from the abdominal portion of the sac into the scrotal part. (See fig. 2.)

The cause of the impulse in this case was evidently the sudden pressure exerted in the act of coughing, upon the upper end of the somewhat flaccid scrotal sac, by which, when the patient was in the recumbent position, the fluid was driven more into the lower end.

The sensation was indeed very similar to the impulse felt in the crural extension of a psoas abscess; and it was clearly, as I pointed out at the time, connected with *fluid* in the sac and not with gut or omentum. The state of things in Case V. was much more simple (fig. 3), being in fact the condition known as strangulation of a hernia *in the sac*; but it is none the less important, from the clinical point of view, because it is not rare.

It is necessary to understand at the outset that it is essential for the prevention of impulse in the hernial tumour that the constriction should be *at least as high up as the abdominal ring*.

If the level of the strangulated point be at any situation below this whilst the abdominal rings are free, all that portion of the gut forming the part of the rupture above the constriction will afford the ordinary hernial impulse, for the simple reason that this gut in the upper part is *not strangulated*. It follows, therefore, that the lower the seat of strangulation is situated in the sac, the greater the quantity of the unstrangulated contents is, and the freer and stronger will the impulse be.

Should the point of strangulation be very high up near the rings, it would necessarily be hardly possible to avoid the diagnosis of a strangulated hernia, excepting from almost criminal carelessness or ignorance; but the matter is by no means so easy when a small knuckle of bowel is nipped, as in this case, in the extreme lower end of a long sac; the amount of unstrangulated gut being altogether in excess veils the small piece which is strangulated. Under such circumstances the small knuckle involved in the stricture may escape notice, especially if, at the same time, as not infrequently happens, the tunica vaginalis is the seat of an old hydrocele. A mistake of this kind, however, ought hardly to be possible; and at all events, if the diagnosis is by any chance not clear, there is no room for doubt as to the proper treatment. The diagnosis, in reality, should not in any case be difficult; for if the symptoms of obstruction depend upon strangulation in the hernia, it will invariably be found that, *although the impulse is present, there has been some recent change in the condition of the hernial tumour.*

This change may take the form of increase in size, or the rupture from having been always reducible may become irreducible; pain is also generally present in the tumour when the actual strangulation occurs; but this is often thought by the patient hardly worthy of mention, because it seems insignificant, especially in advanced cases, when compared with the more distressing pain about the belly and navel, which sets in later in the course of the case.

If, therefore, symptoms of obstruction develop in a patient who has a rupture, and there is any history of some change in the hernial tumour, such, for instance, as those just mentioned, it may be with certainty concluded that the hernia is strangulated, and herniotomy should be at once performed, even if impulse on coughing is present in the tumour.

Indeed, in any case of intestinal obstruction in a patient who is the subject of rupture (this is the old surgical teaching, which is above criticism, and should always be followed), the first indication is to explore the hernia, whether it appears to be strangulated or not.

To avoid any misapprehension on the subject

under discussion, it is now necessary to call attention to another point of some clinical importance.

Although strangulation in the sac of a hernia is commonly associated with the presence of impulse on coughing, it is not *necessarily* so, for *although strangulation may exist in the sac the hernial impulse may be entirely absent*, because the hernia may be *strangulated at the abdominal ring as well as in the sac*, in which case the strangulation occurs primarily at the abdominal ring, the constriction in the sac being caused by the alteration in the bulk of the gut (due to the distension consequent on the primary strangulation), which produces serious changes in the relation of the coils of intestine to omental bands, sacs, &c.

Here is a typical example of a very unusual kind :—

CASE VI.—*Intestinal obstruction of six days' duration from strangulation of a scrotal hernia* (1) *at the abdominal ring, and* (2) *in the sac, by a Meckel's diverticulum; herniotomy; death from senile exhaustion.*

A Chelsea pensioner, 78 years old, was admitted into the Oxford Ward on February 13, 1891. He had been ruptured for forty years; at times he had worn a truss, but for five years no truss had been used, as for that period the hernia had been irreducible. On the night of February 7 the rupture became painful during an attack of coughing, and from that time vomiting occurred several times daily.

On the day before he came to the hospital the pain became much more severe, and the vomiting increased in frequency and severity. The last action of the bowels was on the 9th, but there was continual severe straining, resulting in the passing of a very little wind occasionally.

When admitted the patient was very feeble; the tongue was hard and dry; he was in great pain, and there was a constant forcing cough from bronchitis, which, he said, 'tore him to pieces.'

The left groin and scrotum were occupied by a very large swelling, tense, not tender, without impulse, but tympanitic on percussion. The abdomen was neither distended nor tender.

Chloroform having been administered, I at once performed herniotomy. The sac contained about 12 feet of the small intestine arranged spirally, coil upon coil, which were firmly united by old fibrous adhesions too extensive to admit of any attempt at separation; about 1 foot of the colon, the cæcum, and appendix cæci were also in the sac, but free of adhesions, evidently forming the most recent part of the hernia. There was a tight stricture at the abdominal ring, and in the sac a piece of the small gut, a short distance from the ileo-cæcal valve, was tightly nipped by a fleshy-looking band which stretched across it and was tightly adherent by its end to the sac-wall. The stricture at the ring having been divided, the intestine was disentangled as much as possible. On separating the adhesion which connected the fleshy strangulating band crossing the ileum, this was found to be a pouch-like diverticulum from the bowel, which was before its separation so stretched that it was not recognisable. In spite of the division of as many adhesions as seemed possible, the coils could

*D

not be unravelled sufficiently to admit of the return of the mass of bowel into the belly without slitting up the abdominal wall for about four inches, when, after a little difficulty, the whole hernia was put back. The sac was obliterated, the margins of the opening into the belly firmly closed with silk suture as in abdominal section, the edges of the superficial wound united with silkworm gut, and the scrotal portion of the sac drained.

The patient lived four days in comfort: no vomiting followed the operation, the bowels acted spontaneously on the 12th. He died on the following day, partly from senile asthenia, and partly from exhaustion caused by the continual rasping cough.

Post-mortem examination: ' Ten feet from the cæcum ' is a mass of adherent coils of small intestine, with ' marks of constriction in the lower part; the adhesions ' are old and are in places pigmented. In the middle ' of this mass is a diverticulum lying in contact with ' the mesentery. There is no gangrene; slight dulness ' of the peritoneal coat, but no recent peritonitis. The ' mucous coat of the bowel is normal.

' Kidneys granular and cystic. Lungs very emphy- ' sematous. Heart fatty.'

In this patient it was evident that the coils of the intestine, firmly matted together, and lying tightly adherent to the sac, had for years performed their functions comfortably and effectually until the constriction of the neck of the hernia led to the double strangulation. The case is interesting as an example of the remarkable complications which may be met with in such circumstances.

An important question arises with respect to the treatment of these cases, in which large masses of matted intestine, also adherent to the sac, are found upon performing herniotomy ; for it may well be asked why, as these adherent coils had done their work so well for so long a time, I was not content merely with free division of the constriction (thus relieving the strangulation), instead of subjecting the patient to the relatively severe operation of separating the gut from the sac, and laying open the abdominal wall for several inches in order to return these matted coils into the belly; especially as it was open to some doubt (very slight, it is true) whether they would act as efficiently when loose in the abdominal cavity as when lying in the sac, supported by adhesions.

I was led to adopt this apparently severe measure because the inveterate bronchitis from which this old man suffered was associated with a continual cough of the hardest and most straining kind, which, if I had been content with the mere division of the strangulating ring, which must have been very free to be effectual, would certainly, in my opinion, have forced more and more gut down into the sac. The result of this would have been to cause further strangulation, if nothing worse, for the whole scrotal wound may have been forced open. I therefore considered that the only rational treatment was to enlarge the wound in the manner described, return the whole mass, and obliterate the entire wound and ring by stout sutures which, for the time being at all events, were certain to prevent any further descent of bowel.

The chances of the adherent coils becoming obstructed in their new position seemed to me so slight as to hardly affect the question.

The treatment was fully justified by the result, for the circulation through the matted coils was free, as was shown by the action of the bowels on the third day after the operation, the healthy condition of the parts involved in the operation being sufficiently demonstrated by the *post-mortem* examination.

In cases like this I am sure that the plan adopted is in accord with the principles of sound surgery, and is also practically good. In my own practice I therefore make it a rule always to get the bowel back into the abdomen, even if the adhesion to the sac is so firm as to make it necessary to dissect away a portion of the sac, and return it with the gut still adhering to it. The return of large pieces of the sac still adherent to the bowel appears to add nothing to the risk of the operation, nor does it invalidate the final result. In a case, for instance, of very troublesome irreducible hernia, upon which I operated with a view to the ' radical cure,' it was necessary, in order to return the bowel into the abdomen, to dissect off and leave attached to the gut a strip of the sac more than three inches long and nearly an inch wide. The case pursued the ordinary course, and no drawback of any kind resulted from the condition of the returned parts. The danger resulting from leaving the gut in the sac in these patients with forcing coughs is no imaginary one. I well remember some years ago being asked by a distinguished surgeon to sleep in the house of a patient who had been operated on for strangulated inguinal hernia ; the bowel was so adherent that it was thought undesirable to attempt its return ; free division of the ring only was therefore practised, and the bowel left in the sac.

During the first night there was much cough, and the tumour increased in size, with considerable pain ; on the following morning, during a violent attack of coughing, acute pain was felt in the rupture, and vomiting occurred ; on the removal of the dressings soon afterwards a large quantity of intestine lay *beneath the dressings on the patient's thigh*, the stitches uniting the wound edges had given way under pressure, and the gut had escaped from the sac through the gaping wound. It is needless to say that the patient, an oldish man, did not long survive.

This case made a great impression upon me at the time, so much so that for my own part I will on no account leave gut in the sac of a hernia under these circumstances, however adherent it may be, unless its return, from some extraordinary reason, is altogether impossible.

So important do I consider it that no gut should be left in the sac after herniotomy in cases like the above, that I should not hesitate to consider the propriety of resecting the adherent intestine, which I think would be justifiable if the obstacles to reduction were due to local causes only (*i.e.* adhesions, &c.), and if the age of the patient, and the other conditions generally, afforded a reasonable prospect of the operation being successfully performed.

LECTURE III

ON SYMPTOMS OF STRANGULATION OCCURRING IN CASES IN WHICH THE HERNIAL SAC APPARENTLY CONTAINS NEITHER OMENTUM NOR BOWEL

(Not previously published)

SYNOPSIS—Additional points of interest suggested by the cases described in Lecture II.—THE OCCURRENCE OF SYMPTOMS OF STRANGULATION WHEN THE SAC CONTAINS NEITHER BOWEL NOR OMENTUM— Further reference to Case IV.—Clinical example (No. VII.), symptoms of strangulation in a woman occurring in connection with an inguinal sac shut off from the abdominal cavity, which contained fluid only—Comments on this case—Sac formed by unobliterated extremity of canal of Nuck—Clinical example (No. VIII.), intestinal obstruction in a case of supposed femoral hernia in which the sac contained fluid only, and freely communicated with the abdominal cavity—Comparison of these two cases—Three classes of case upon which neither gut nor omentum is found in the sac after herniotomy in cases of strangulation—(1) Sacs completely shut off from the abdomen—(2) Sacs having free communication with the abdominal cavity—(3) Sacs which at first appear to be cut off from the belly, but really communicate—Reasons for occurrence of symptoms in these conditions discussed—Clinical example (No. IX.), irreducible painful femoral hernia in which the sac, although apparently occupied only by fluid, contained a small nodule of nipped omentum—Remarks on this case—Reference to a case showing the possible influence of adherent omental bands inside the belly in causing signs of apparent strangulation—Practical application of this point to treatment.

APPENDIX—Clinical example (No. X.), acute intestinal obstruction apparently connected with hernia of extra-peritoneal tissue—Operation —Released Richter's hernia found at *post-mortem* examination— Remarks on this case—Reference to similar instance recorded by Mr. Gay—Practical bearing of Case X. upon explanation of symptoms in Case III.

IN addition to the direct clinical lessons already discussed, the following points, which are of much practical importance, arise in connection with the cases described in the last lecture :—

1. The occurrence of symptoms identical with those of strangulation of hernia in cases in which the sac contains neither bowel nor omentum, and in which herniotomy completely relieves the symptoms.

2. The occasional existence in the sac of membranous flaps possessing a valve-like action.

3. The peculiarities met with in the shape of hernial sacs, and the causes of these deviations from the ordinary type.

In the present lecture I propose to consider the first only of these interesting points.

The Occurrence of Symptoms of Strangulated Hernia in Cases in which the Sac contains neither Gut nor Omentum.

In the description of Case IV. it will be seen that upon opening the scrotal sac about two ounces of fluid escaped, and that no communication between this sac and the abdominal cavity was at once made out, although subsequently a free opening was found. At first it was not unnatural to conclude that this was an instance of the coexistence of the symptoms of strangulation with a sac containing fluid only, although the semi-flaccid state of the parts before incision was entirely opposed to that supposition.

The dull area in the abdomen, however, of course led to further careful investigation, with the result that my finger passed behind the margin of the valve-like fold into the abdominal sac.

This, therefore, was not an example of the kind of case I am now dealing with, because, although the scrotal sac contained only fluid, it was directly continuous with the abdominal portion which contained a strangulated knuckle of gut.

The following case, recently under my care, is an excellent illustration of the coexistence of symptoms of strangulation and a sac containing fluid only.

CASE VII.—*Tense irreducible inguinal tumour in a woman ; symptoms of strangulated hernia twice recurrent ; herniotomy ; sac containing fluid only ; immediate and complete relief.*

M. F——, a laundry-maid, aged 33, was admitted into Princess Ward on September 30, 1891. She had been ruptured for fourteen years, and had always worn a truss.

In 1881 she was a patient in St. Bartholomew's Hospital, on account of 'sickness and stoppage of the ' bowels,' and the rupture ' was put back whilst she ' was under chloroform.' From that time the hernia had been reducible, and at night when she lay down it disappeared of its own accord, until three or four months before her admission into St. George's, when its reduction became more difficult and could be only gradually effected, whereas before that time it could always be 'put back all at once.'

A fortnight before she came under my care she found that the hernia could not be returned at all. Pain and vomiting soon came on, and great constipation followed, requiring very strong purges, which caused only a slight evacuation with great pain.

On lying up the pain decreased somewhat, and the sickness subsided; the constipation, with the irresistible straining, however, continued. Upon getting about again the same symptoms occurred, and she then noticed than when the pain was more severe the tumour increased in size and became tender; thereupon vomiting invariably followed.

On admission.—A healthy-looking woman; tongue very thickly coated, breath rather offensive.

In the right groin, evidently passing down from the inguinal canal, was a tense rounded tumour, as large as a hen's egg, which extended into the upper end of the labium, irreducible, without impulse, but tense and tender.

There was pain over the lower part of the abdomen, running up to the navel, which was increased by a deep breath or violent coughing.

The belly was neither distended nor tender.

Herniotomy was performed, and a thick sac exposed, which upon being laid open was found to contain a little fluid and nothing else, with the exception of two small old blood cysts attached to its wall. There was no communication of any kind with the abdominal cavity, the approximated sides of the neck of the sac having become completely adherent, and thus the canal was entirely obliterated. The sac was ligatured and removed, the superficial parts being then closed in the ordinary way with silkworm gut sutures. Every symptom was immediately relieved by the operation; she left the hospital on Oct. 21, and after a fortnight's stay at the Convalescent Hospital she returned to her employment. The patient was heard of again in the middle of December 1891; there had been no recurrence of any of the symptoms, the bowels had been acting with regularity, and she was doing her work, which was very heavy, with perfect comfort.

This may be taken as a fair illustration of the manner in which a distended peritoneal sac, which at one time had no doubt contained a hernia, may set up symptoms resembling those of strangulation of the hernia in the absence of any communication with the cavity of the peritoneum. It is clear from the details of the case that the obliteration of the neck of the sac was comparatively recent, and also that the closure had been gradually effected, as shown by the progressive slowly increasing difficulty which was experienced in returning the contents of the sac into the belly. In order that there should be no doubt as to the peritoneal nature of this cyst, the portion removed was submitted to Dr. Delépine with no other particulars beyond the fact that it had been taken from the groin of a female patient. The following is his report :—

'In the walls of the cyst on one side there are
' many bundles of unstriped muscular fibres and some
' pretty large vessels and nerves (round ligament).
' Considering the situation, it is evident that the cyst
' is probably the blunt extremity of the canal of Nuck
' separated from the rest of the canal by constriction
' (as the tunica vaginalis in the male).'

The fact that this sac was composed of the end of the patent canal of Nuck to some extent no doubt explains its tendency to closure, which was obviously far greater than is usually seen in hernial sacs (excepting those formed by the funicular process in the male), at the time of life reached by this patient.

The symptoms here had not quite attained the acute stage, but there is no doubt that had the girl continued her work the consequent increased tension in the cyst would have been followed by all the symptoms of acute intestinal obstruction.

The following case well illustrates the occurrence of acute symptoms in connection with a sac containing neither gut nor omentum, and which had a free communication with the belly.

CASE VIII.—*Symptoms of acute intestinal obstruction in a woman who had a femoral hernia; herniotomy; sac containing a little fluid only; complete and immediate relief with ultimate recovery.*

C. D——, a laundress, aged 46, was admitted into Princess Ward on October 20, 1879, under the care of Mr. Holmes. She had been ruptured on the right side for twenty years. A truss had never been worn. The hernia had always been reducible until a week before her admission, when she felt suddenly, during straining at stool, some very sharp pain in the groin.

With some difficulty she returned the greater part of the rupture, but could not put back the whole of it. The pain increased, and three days before she came to the hospital the rupture again increased in size and could not be reduced at all, either by the patient herself or by a medical man who was called in.

For two days there had been much griping, and great nausea without actual vomiting. The bowels acted for the last time thirty-six hours before admission, the action previous to that having been nearly a week before.

On admission the patient was somewhat feeble, and a little anxious in aspect. The tongue was white and thickly coated; pulse, 80.

She complained of much griping pain over the lower part of the abdomen, which was rather tender but not markedly distended.

There was extreme nausea, but no vomiting.

In the right groin, just below Poupart's Ligament, was a nodulated tumour which was pretty clearly a femoral hernia, about as large as a pigeon's egg, without any impulse whatever.

She was placed in bed and an ice-bag applied. On the following morning the nausea had subsided and the pain was less, but in the afternoon (4.45 P.M.), she vomited and the pain increased. At 8 P.M. vomiting again came on, with further increase of pain.

Mr. Holmes therefore performed herniotomy (I happened to be present at the operation). A small hernial sac was exposed, the walls of which seemed very thick. The sac contained a little peritoneal fluid and nothing else. There was an opening through the neck of the sac large enough to admit the tip of the index finger into the abdomen. The neck of the sac was obliterated by means of silver suture.

Immediately after the operation a copious passage of flatus occurred per anum, the first since her admission.

The operation entirely relieved the symptoms; the bowels acted regularly, and with the exception of a little irritation about the wound the patient made an uninterrupted recovery. She was seen eight months later, never having had a recurrence of discomfort of any kind.

This case differs from the one which precedes it in two particulars, for it was of the femoral variety, and there was a free communication between the sac and the abdomen.

There appears, therefore, to be three distinct classes of case in which, upon performing herniotomy for the relief of apparent intestinal obstruction, the sac is found to contain neither gut nor omentum, and may indeed contain nothing, not even fluid.

These varieties are as follows :—

1. Sacs having no communication with the peritoneal cavity, and which are always distended with fluid. (Example, Case VII.).

2. Sacs having a free communication with the abdomen, and from which a little fluid may or may not escape when they are laid open. (Example, Case VIII.).

3. Sacs which at first sight appear to be completely shut off from the peritoneal cavity, but which are in reality only temporarily so shut off by the pressure which is exerted by accumulations of fluid upon membranous flaps in the sac, which have a valve-like action. (Example, Case IV.).

Sacs of femoral herniæ which have become cut off from the abdominal cavity, either by general constriction and obliteration of the neck or by membranous diaphragms, are rarely if ever seen ; but in the infantile form of hernia (*i.e.* into the funicular process), and also, but less commonly, in the congenital variety, peritoneal cysts having no communication with the general peritoneal cavity, although continuous with the peritoneum, are by no means infrequent. Whether in all cases these sacs or cysts have contained herniæ is a matter which may be open to question, but that when distended they give rise to all the signs of strangulation of gut or omentum there is no doubt.

What is the explanation of the occurrence of these apparently serious symptoms when no strangulation of gut or omentum seems to exist, and why does the mere opening of these sacs immediately and completely relieve the symptoms ?

The reasons for the symptoms of strangulation, so far as I understand the matter, are somewhat different in the three conditions referred to.

In cases in which there is no communication between the sac and the abdominal cavity, or in which the opening is temporarily obliterated by a valve-like membranous flap, the symptoms are caused, I believe, by the mere distension of the sac, which is. of course, still *continuous* with the general peritoneum.

This distension, either by simply dragging upon the peritoneum, or by reflex causes, excites so much intra-abdominal irritation as to produce the vomiting and constipation which so closely simulate intestinal obstruction or peritonitis ; such irritation, in fact, as a ligature placed tightly around a piece of omentum not infrequently causes in cases of omental hernia after operation. This irritation would theoretically be more likely to occur in a situation like the groin, where the nerves in relation with the sac are both in their numbers and in the intimacy of their connection with it in excess of those in any other locality.

Practically, it also appears that this is the case.

When a free opening exists between the sac and the abdominal cavity, the symptoms are in the majority of cases due, I believe, to the actual strangulation of a small piece of omentum or gut which has become engaged in the ring. The manipulation of the sac after its exposure, and the incision subsequently made into it, so alters the tension of the parts about the neck, that the small strangulated mass slips back of its own accord into the belly and escapes notice altogether.

That this is possible the following case shows :—

CASE IX. — *Irreducible painful femoral hernia ; herniotomy ; sac-contents 1 oz. of fluid and a small nodule of omentum which, whilst the sac was being held up after incision, slipped back spontaneously into the abdomen.*

A woman of middle age was under my care in the Princess Ward. For many years she had suffered from a femoral hernia in each groin. A truss had occasionally been worn, but had not been used with regularity as it appeared to afford no comfort.

Both herniæ had been reducible until a few months before she came to the hospital, when that on the left side became suddenly somewhat larger than usual and rather painful. From that time the left rupture had been irreducible, did not disappear when she lay down as it had always done before, and, at the end of a day's work, had always become tender and very hard.

At times, when the tumour was larger and more tender than usual, there was pain around the navel and nausea was felt. On one occasion only had she vomited.

On admission the patient was healthy-looking and seemed well.

In the right groin was a small reducible femoral hernia. On the left side, over the saphenous opening, was a rather tense irreducible swelling, with slight but distinct impulse on coughing. It was neither tender nor painful, but the attacks of pain, &c., when she was up and about were so frequent that she was anxious for relief by operation. I therefore operated with a view to performing the 'radical cure.'

On opening the sac, which was not larger than a walnut, it appeared at first to contain only a little fluid, but on holding up its walls in order to get a better view of its interior, there was seen occupying the neck, like a cork, a small piece of omentum about the size of a horse-bean which, whilst I was on the point of picking up a pair of forceps with which to seize it, for the application of a ligature, all at once slipped back into the abdomen and disappeared from view, leaving a perfectly empty sac. On passing my finger through the neck of the sac into the belly, the small nodule could be felt, so I hooked it down, ligatured it, and cut it off in

E

order to prevent any chance of its becoming engaged in the ring again.

The sac was afterwards obliterated at its neck by suture and removed in the usual manner. An uninterrupted recovery followed, and no further discomfort of any kind was subsequently felt by the patient.

In this case I have no doubt, seeing the very loose manner in which this small nodule lay in the neck of the sac, that any decided manipulation of the sac before it was laid open might easily have led to the spontaneous withdrawal of the little herniated mass into the belly; and so upon the sac being laid open it would have been found to contain no hernia at all, and probably not even any fluid, as the small amount which was present would have flowed back into the general peritoneal cavity.

Altogether this case seems singularly à *propos* to the matter under discussion.

There appears to be yet another possible cause for these cases, judging from the following condition, found by chance in a subject in the *post-mortem* room. The subject was a male, who had a scrotal sac of small size in which no hernia lay when it came under observation. Passing down from the transverse colon, and causing quite an abrupt bend in its course, was a long and thin omental band, firmly adherent below to the peritoneum immediately inside the right internal abdominal ring. It was evident that if the sac had been occupied by a hernia or fluid whilst the subject was in an upright position, the weight in the sac would have drawn the end of this band considerably down, not only thus making additional traction on the colon, but also bringing the end of the adhesion well into the neck of the sac, and so rendering it liable to irritation from becoming nipped by any constriction about the ring.

This condition is clinically important, because it shows the necessity in any case where the sac contains no gut or omentum, and yet the symptoms of strangulation exist, of making a careful digital examination inside the abdomen, when the neck of the sac allows of the introduction of the finger : not only with a view to the discovery of any internal strangulation, but also to ascertain whether any band or adhesion is attached around the abdominal aspect of the ring in such a way that under certain circumstances it may be dragged upon, or possibly nipped, sufficiently to produce symptoms of an apparently serious kind.

APPENDIX TO LECTURE III

THE following instance, which came under my treatment whilst these pages were being prepared for the
press, illustrates very forcibly the difficulty in deter-
mining the cause of the symptoms of intestinal
obstruction in some of these anomalous cases :—

CASE X.—*Acute intestinal obstruction in a man aged
73; irreducible tumour in groin, which upon operation
proved to be subperitoneal tissue nipped in femoral ring;
no hernia or hernial sac discovered; complete relief of
symptoms; death from senile asthenia; partial enterocele recently released from strangulation found at* post-
mortem *examination.*

Wm. L——, aged 73, was admitted under the care
of my colleague, Dr. Ewart, on December 19, on account
of acute intestinal obstruction which supervened suddenly four days previously.

He had for many years been very constipated, and
sometimes had attacks of griping pains in the right groin.
He was not aware that he had ever suffered from rupture.

The bowels acted for the last time, prior to admission,
on the 15th, and vomiting commenced after an attack
of the griping pain on the 16th, and afterwards frequently
recurred.

On admission, the patient was much collapsed and
extremely feeble. There was vomiting of stinking dark-
coloured material. The radial and temporal arteries
were so hard and rigid from atheroma that the usual
pulsation could not be felt.

The abdomen was neither distended nor tender.

There was pain extending from the right groin up to the region of the liver. At intervals snake-like movements of the small intestine, well seen through the parietes, occurred with much spasmodic pain.

In the right groin was an oval mass about an inch in length, lying in the saphenous opening, and presenting all the characteristics of a strangulated omental femoral rupture.

I was, therefore, requested to see the patient, and at once determined to perform herniotomy in the usual manner.

On exposing the tumour it was found to be a mass of tissue and of conical form, not larger than a sparrow's egg, which had all the appearances of old strangulated omentum. Above by a sort of stalk it passed through the femoral ring, the margins of which nipped it tightly. The ring was freely divided as in herniotomy, and the mass when fully liberated in this way was drawn down. It was then seen to be attached to the peritoneum, a portion of which was easily pulled down through the ring by dragging upon the little mass. In order to be sure that no internal strangulation existed, I opened the peritoneum thus brought into view, and introduced the index finger into the belly, but found nothing abnormal. A ligature of catgut having been applied around the peritoneum above the opening which had been made, the distal part with the mass attached was cut away.

The superficial wound was then brought together in the ordinary way.

All the symptoms were relieved, no further vomiting occurred, no more pain was felt, and the bowels acted naturally within twelve hours.

Unfortunately the patient was too old and feeble to rally enough to recover, and died of senile asthenia.

Post-mortem examination : ' Sixty-five inches from the

' cæcum the ileum is very deeply congested, and has the
' serous coat absent from an area the size of a three-
' penny bit, as if it had been the seat of a partial
' enterocele. The bowel is very weak, and easily rup-
' tured by a little manipulation. Around the area of
' the partial enterocele the intestine appears narrowed.
' Above this point the intestine is dilated, and contains
' liquid fæces; below it is collapsed, but only to a slight
' degree. No peritonitis, no band, or other cause for con-
' striction of ileum found.'

This case, although not of quite the same nature
as those discussed in the foregoing lecture, is of great
interest when considered in connection with them,
as no evidence of a hernial sac nor hernia of gut
or omentum was found at the operation, the pyra-
midal mass being entirely extra-peritoneal and lying
merely in a capsule of cellular tissue.

If the patient had recovered, as he certainly
would have done had he not been so feeble and
atheromatous, I should naturally have concluded
that the cause of the obstruction was in some way
directly connected with the strangulation of the mass
of extra-peritoneal tissue, but the co-existence of a
strangulated knuckle of gut would hardly, under the
circumstances, have suggested itself. The recently-
strangulated bowel found at the *post-mortem* examination was undoubtedly liberated either by the
dragging down of the peritoneum, for purposes of examination, or by the finger when introduced through
the small opening made subsequently. I was, how-

ever, not conscious of having reduced or pushed away any gut from the ring with the finger end. The pouch of peritoneum, to which the little mass was attached, and which I opened, was in fact a small hernial sac lying at or just inside the femoral ring.

The clinical importance of the case is obvious, as it shows the necessity of a thorough examination of the parts under such conditions. Had I been content with the mere division of the stricture and the removal of the little mass without further investigation, it is almost—indeed, I suppose quite—certain that the strangulated gut would have remained unreleased.

Cases like this are rare, and I myself have never had to deal with an exactly similar one. Mr. Gay, however, records an instance very like it in Vol. XXIV. of the 'Pathological Society's Trans-' actions.'[1] Case III., described in Lecture I., in which intestinal obstruction occurred in connection with an abscess in the crural canal without any apparent hernia, is of additional interest when considered in the light afforded by this instance, for it is quite possible there may have been a small hernia at the femoral ring, which was released from strangulation by the manipulation to which the parts were subjected in treating the abscess.

[1] A case of enteric obstruction, with a rare form of femoral hernia ; operation ; death.

LECTURE IV

ON SOME PECULIARITIES MET WITH IN THE AR-RANGEMENT AND SHAPE OF HERNIAL SACS

(Not previously published)

SYNOPSIS—OCCASIONAL OCCURRENCE OF MEMBRANOUS FLAPS WITH A VALVE-LIKE ACTION IN THE NECKS OF HERNIAL SACS—Further reference to Case IV., in which a valve was present preventing passage of fluid from abdomen into scrotum. Clinical example (No. XI.), valve-like flap obstructing entrance into abdomen from scrotal sac in a child—Remarks on these flaps—Absence of literature on the subject—Reference to case described by Mr. South—Obstacles to reduction of herniæ presented by such flaps even after division of stricture—Importance of dividing these flaps in herniotomy, even if no apparent pressure on the gut is produced by them—Reason for this treatment—History of the valve-like flaps—Sir Astley Cooper's opinion—Author's views and reason for the same—Clinical importance of the flaps—Reference to cases in evidence of this, especially with regard to treatment.

PECULIARITIES IN THE SHAPES OF HERNIAL SACS AND THEIR CAUSES—Typical form of sac—Simplest deviations from this—Variation in degree—Hour-glass sac—Multilocular sac—Modification of hour-glass form peculiar to inguinal hernia produced by diaphragms in sac—Irregularities in sac before exposure by dissection often disappear after the sac has been exposed, especially in some inguino-scrotal cases—Contents of very irregular sacs, usually both gut and omentum—Irregular or hour-glass forms frequent causes of strangulation in the sac—Strangulation at abdominal ring more serious in irregular than simple sacs—Reasons for this—Causes of these deviations from typical form—Alteration in form involving changes in the sac-wall itself—Mode of production—Reference to old theory of spontaneous rupture of sac—Mr. South's case—Comparative rarity of very irregular sacs in inguinal hernia explained—Irregular sac caused by co-existence of oblique and direct inguinal rupture—Multiple sacs in umbilical hernia and their practical bearing on treatment—Modified hour-glass sac in inguinal hernia, and its causes.

1 now propose to discuss the second and third points of interest alluded to at the commencement of the last lecture.

The occasional Existence in the Neck of the Hernial Sac of Membranous Flaps having a Valve-like Action

In Case IV., described at page 20, there was present at the point of communication between the abdominal and scrotal portions of the sac, a membranous flap, suspended from the anterior and lateral aspects of the sac wall at its constricted part, which was so arranged that under the increased pressure caused by rapid effusion into the upper part of the sac it acted as a valve, preventing any passage of fluid from above into the scrotal pouch.   Nevertheless, upon raising this flap with the end of the finger, a free passage was found between the two portions of the sac.

The following is an interesting case, which shows that a membranous flap may act in the opposite direction :—

CASE XI.—*Irreducible translucent scrotal tumour in a child who had suffered from strangulated hernia; operation; hernial sac laid open—valve-like flap at upper end of sac preventing passage of fluid into abdomen, although a free communication existed.*

R. M——, a boy, aged 2 years, was admitted into the Princess Ward under my care on Sept. 29, 1891.

The child had been ruptured from birth.   Four

months before he was brought to the hospital he had
been operated on for a strangulated hernia which was
reduced without opening the sac.   Very soon the scrotal
tumour reappeared, gradually increased in size, and could
not be returned.

On admission.—The child was healthy-looking in
aspect and seemed well generally in every way.   The
right side of the scrotum was occupied by an elongated,
rather tense swelling, entirely without impulse, fluctuating, translucent, and quite irreducible.

The testicle was natural, and lay separate from the
swelling quite at its lower end.   Operation was performed with a view to the 'radical cure.'   On the tumour
being laid open, it was found to be a peritoneal sac
containing nothing but clear fluid.   The sac passed
down to the testicle, but did not communicate with the
tunica vaginalis.   Upon examining the neck of this sac
there at first seemed to be no communication with the
abdominal cavity, as the entrance of the finger was
arrested by a membranous septum.   Subsequently there
was found on passing the finger along the neck of the
sac, a lunated edge beneath which, after it had been
raised up, the finger readily entered the belly through
an opening at least half an inch in diameter, represent-
ing in fact the entire calibre of the sac neck.   The
'radical cure' was performed by ligation and inversion
of the sac, the margin of the ring being brought together with tendon sutures.   The boy left the hospital
well on October 7.

In this case it was clear that fluid could pass
easily from the abdomen into the scrotal sac, but
could not be returned in the opposite direction in
consequence of the existence of this valve-like flap.

The sac was undoubtedly that of a hernia into the funicular process of the peritoneum.

These are the only two instances with which I have met in which such *complete and competent* valves exist in any part of a hernial sac, and I do not find any specific reference to such complete examples in the literature of hernia excepting by Mr. South, who, in his translation of Chelius (vol. ii., p. 77), describes an operation upon a large, irregular, strangulated hernia, as follows :—

' Having introduced my finger into the sac ' (after it had been laid open by the knife), ' I could ' not at first pass it down to the stricture, as it was ' intercepted by a band which I supposed to be an ' old adhesion ; but having drawn the omentum and ' gut to the outer side, I was enabled to reach, and ' found the stricture very tight, and admitting only ' the tip of the finger, but sufficient to allow the ' entrance of the blunt-ended bistoury, which I divided ' until my finger would pass into the belly up to the ' second joint. I then readily emptied the gut, and ' attempted to return, but could not succeed. It was ' thought that the difficulty depended on the stricture ' not having been sufficiently freed, and I therefore ' prepared again to introduce the bistoury by draw- ' ing the omentum and intestine to the outer side. ' This, however, having been done, *a broad membrane* ' *was seen descending from the upper part of the sac,*

' behind which the finger could be passed.   It was this,
' doubtless, which first prevented the introduction
' of my finger into the stricture, and subsequently
' obstructed the entrance of the gut into the belly
' by dropping against the mouth of the sac.   We
' determined on its division, and this done *with-*
' *out further dilatation of the stricture,* the intestine
' easily returned.'[1]

This case of Mr. South's is clearly one of a
similar kind to the cases I have described here.
Valve-like membranous flaps of this sort occur, I
have no doubt, much more commonly than is generally supposed; for although, as already mentioned,
I have seen only these two instances in which the
valve-like action has been so perfect as to prevent
the passage of fluid, I have not unfrequently found
in strangulated inguinal herniæ one or more membranous flaps in the neck of the sac which, whilst
they have *appeared* to exercise no pressure at all
upon the bowel, undoubtedly presented so much
obstacle to its return, even after free division of the
stricture, that the reduction could only be safely
effected after the division of the flaps.

In any case, therefore, in which such flaps exist,
whether they be small or large, and even if they seem
to exert no pressure upon the gut, I invariably divide

----

[1] The italics are mine.

them freely before proceeding to return the contents of the sac.

This practice is, I believe, good, and may with advantage be followed in all such cases, for if these flaps or folds do not prevent the return of the gut altogether, they sometimes offer sufficient resistance to place the bowel in danger of being seriously bruised, and may perhaps lead to actual splitting of its peritoneal coat, if attempts at reduction be too long persisted in, or if too much force be used before they have been divided. A further reference to this point will be found in Lecture V.

With respect to the history of these valve-like flaps I am not clear, nor can I find any decided explanation of their formation. Sir Astley Cooper, in his work on Hernia, speaks of septa forming in the mouth of the sac, and appears to think that they are produced by the pressure of the parts around, but nothing so complete and membranous as these flaps which I am discussing could be so formed. He also says : 'A membranous band crossing the mouth of ' a hernial sac might produce strangulation by its ' pressure upon the intestine,' but he does not mention that he ever met with such a case. Moreover, this remark refers rather to bands of adhesions, inasmuch as he goes on to say that they 'appear to be produced ' in the following manner. During the reducible ' state of the hernia inflammation takes place in the

' contained parts and in the inner surface of the sac,
' but by using proper means the protruded parts are .
' reduced and the sides of the sac collapse and adhere
' together.   However, while the adhesions are in
' a glutinous state, a fresh descent takes place from
' the abdomen, and the hernial contents again dis-
' unite the surfaces of the sac everywhere except at
' the points of union of these inflamed parts, the
' cementing lymph of which, instead of bursting
' asunder, elongates with the fresh pressure and forms
' these membranous bands which are seen *passing*
' *from one side of the sac to the other*.   Between these
' the intestine and omentum get entangled.'   It is
plain again here, I think, that this admirable description could hardly apply to such valve-like flaps as
those which were found in my two cases.   At all
events, the valve in the child could not have been
thus produced, nor can I see how that in the man
could have arisen in this manner, for in each instance
the flaps consisted of a well-marked involution of the
sac-wall.

I am therefore led to conclude that these flaps
are processes formed in the natural attempt at the
closure of the funicular process of the peritoneum.
The reasons for this conclusion are mainly the following :—(*a*) These flaps occur so far as I know only in
inguinal hernia, and, moreover, are almost always
limited to cases of congenital and infantile hernia.

In the infantile variety—(*i.e.* hernia into the funicular process), these flaps are most commonly seen, and may be two, three, or even more in number, varying in size from slight elevations to large membranous curtains, which may be crescent-shaped, and therefore involve only a portion of the circumference of the neck of

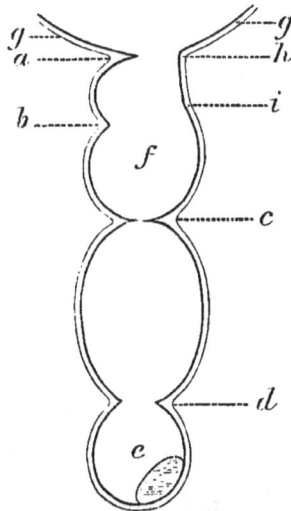

FIG. 4.—DIAGRAM SHOWING ARRANGEMENT OF LONG COMPLICATED SAC FOUND IN A CHILD IN OPERATING FOR THE RADICAL CURE OF A CONGENITAL INGUINAL RUPTURE.

a, *b*, Membranous folds at level of internal and external ring respectively.
c, Diaphragm with pin-hole perforation.
d, Diaphragm with large opening leading into a cyst (*e*), which contained the testicle.
g, Peritoneum passing down to form sac.
h, Level of internal ring.
i, Level of external ring.
The hernia occupied the compartment of the sac marked *f*.

the sac, or may take the form of a diaphragm with a central opening (fig. 4, *c*), in which case they of course involve the whole circumference of the sac, and become a fruitful source of strangulation in the sac. This diaphragm-like form differs from the crescentic folds, in that it occurs in the body of the sac as well as in the neck; whereas I have never seen the crescentic folds excepting in the upper constricted part of the sac—*i.e.*, the neck. (*b*) The fact already mentioned that these flaps are composed almost entirely of a thin involution of the serous membrane. (*c*) Similar folds and diaphragms are occasionally met with in imperfectly-closed funicular processes into which no hernia has descended. In a case, for instance, of congenital hydrocele which I laid open in a young child, a diaphragm existed, at about the level of the external ring, through which an opening not larger than the head of a pin led into a funicular process which freely communicated with the peritoneal cavity above through a very much larger opening. This diaphragm was evidently the result of an unsuccessful natural effort at obliteration. Another mode of production of these folds and diaphragms, more especially the former, seems to me possible, but at the same time so improbable as to make it of little account. Nevertheless it may be mentioned, and is as follows:—In the very quickly increasing and easily reducible herniæ of young subjects the rapid increase in the size of the sac is caused by the dragging down of successive portions of the peritoneum by the traction exerted by the bowel when it passes well down into the hollow of the sac.

With each return of the hernia into the belly this increasing sac becomes shortened again, partly from its own resilience, but principally because the fundus is raised towards the abdominal ring by the action of the cremaster muscle.

The sac remains thus shortened and wrinkled until it is forced down again by the next descent of the hernia.

Under the circumstances, is it not possible that this process of stretching and shortening, repeated again and again, may lead to the formation of a permanent transverse fold at the most fixed part of the sac by the cohesion of the sides of the transverse wrinkles which are produced each time the sac contracts? Thus I venture to think may be developed the occasional, but very rarely seen, flaps met with in the neck of certain acquired herniæ, in which, by the way, the folds appear to be always single.

Whatever may be the process by which these folds arise, there is at all events no doubt that, clinically, they are of great importance; for in cases in which they exist it is not sufficient, as I have said before, to divide the stricture only, but the folds or diaphragms must themselves also be freely cut whether they seem to strangle the gut or not.

In Lecture V. a case of strangulated hernia is referred to in which the peritoneal surface of the gut gave way during an attempt to reduce it after the stricture had been very freely divided, there being no obvious bar to reduction, and the gut

F

itself being in perfectly good condition. No such mishap as this has yet happened to me; but I am sure that it would have done so in one of these cases in which a membranous flap lay in the neck of the sac if I had persisted in my attempts at reducing the intestine. The flap seemed to lie quite loosely upon the bowel, but it was felt, when the finger was placed in contact with it, to grasp the gut during attempts at reduction like a sort of sling, becoming quite loose and flaccid again when the attempts at reduction were discontinued.

The Peculiarities which Occur in the Shape of Hernial Sacs and the Causes of these Deviations from the Ordinary Type.

The typical shape of fully formed hernial sacs generally may be taken as globular or pyriform, the smallest part usually being at the point of communication with the abdominal cavity. The most simple deviation from this typical form is that in which a constriction of the wall of the sac divides it into an upper and lower portion, the size of the communication between the two parts depending upon the degree of constriction. This may be so slight as to be hardly noticeable, or may be so pronounced that the channel between the two portions is not larger than a crowquill. I have myself seen an instance in which the channel admitted with difficulty a No. 7 (English) catheter. This simple peculiarity in shape is, I need hardly say, the so-called 'hour-glass' sac, which may be met with in any variety of hernia, whether it be femoral, inguinal, umbilical, or even ventral.

Very irregular sacs (multilocular) are not uncommon in femoral hernia, are rare in inguinal cases, but are commonest of all in umbilical ruptures. Whilst this very irregular form is rare in inguinal hernia, there occurs in that variety of rupture a kind of sac not met with elsewhere, which may be said to be a modification of the 'hour-glass' shape, inasmuch as the sac is divided into three or more compartments by two or more constrictions or diaphragms, as shown in fig. 4 (page 63).

It is necessary to note that many sacs which appear irregular before they are exposed by dissection, lose the irregularity after the fascia over them has been divided. This is especially well seen in some inguino-scrotal herniae, which often present an apparent constriction just below the upper end of the scrotal part of tumour, which entirely disappears when the sac has been actually exposed, leaving it of the typical shape.

The very irregular sacs almost always contain both bowel and omentum, the latter occupying as a rule the distal portion of the sac, the bowel lying in the proximal part. When the communications between the portions of the 'hour-glass' sacs or their modifications alluded to are very small, the distal part of the sac is not infrequently occupied by fluid only, and the opening leading into it may become closed altogether by adhesion of omentum to its upper margin. These irregular sacs, unless the openings between their several parts are very large, become fruitful causes of *strangulation in the sac*, in consequence of the bowel being forced into the lower compartment through the narrow communication, to which it is often conducted by the stalk of the mass of omentum lying in the distal part. Irregular sacs of all kinds which contain bowel in their distal compartments render strangulation of a hernia at the abdominal ring a more serious condition than it is in simple sacs, in consequence of the tendency of the hernia to become strangulated secondarily *in the sac*, from the increase in bulk and tension which is produced in the contents of the hernial sac by the primary constriction at the abdominal ring, the result being a double strangulation —*i.e.* at the ring and in the sac. (Case VI.)

The manner in which certain symptoms of strangulation are modified by these peculiar sacs is fully discussed in Lecture III.

In considering the causes of these deviations from the simple form of sac, it is necessary to bear in mind that irregularities in the sacs are of two distinct kinds, one consisting of an actual change in shape *in the sac-wall itself* by the formation of pouches, diverticula, or imperfect septa and diaphragms ; the other being a perhaps complicated arrangement of pockets and spaces in the interior of the sac formed by the adhesion of some of the contents of the hernia to each other or to the sac wall, or by some other change in the relation of the contents, such as for instance takes place in the production of omental sacs when the omentum is neither adherent to the sac nor bowel.

Irregular pouch-like arrangements of the sac wall are commonest in femoral hernia, but are sometimes seen in inguinal cases, in which, together with um-bilical cases, a very peculiar form is at times met with.

Excluding for a moment the peculiar form last mentioned, the explanation of the very irregular sac-walls is probably best accounted for by the traditional reason that the irregularity is due to the pressure exerted upon the sac wall by the bands of fascia which cross it, or through which, in the course of its progressive formation, it bursts abruptly or gradually makes its way. Modification of form therefore must necessarily occur in the majority of large femoral herniæ from the pressure exerted on the sac as it turns around the margin of the saphenous opening ; in fact, in some cases quite a ' kink ' is formed in the sac at this spot. The disposition of irregular strands of fascia about Poupart's ligament again must necessarily cause complicated irregularities when the sac grows to any considerable size. When comparatively recent, these irregularities disappear upon division of the constricting band of fascia, but when of long standing the sac, even after complete removal by dissection, often retains permanently the change in shape. The deviation produced by the presence of strands of fascia is not frequently seen in any marked degree in inguinal cases, although, as already mentioned, a slight transverse depression in the sac wall is not uncommon at the point of junction of the groin and scrotum. Occasionally, however, the inguino-scrotal sac becomes irregular, one portion passing down to the scrotum, the other part running up along Poupart's ligament; the depression between the two parts being apparently due in the first instance to a constricting band of fascia. This form appears to be almost confined to inguinal hernia of the *direct* kind.

It was formerly supposed that these very irregular sacs, especially when inguinal, were the result of spontaneous rupture or bursting of the original sac, which was followed by the escape of the sac-contents into the surrounding soft parts, in which secondary spaces were formed communicating, of course, with the true primary peritoneal sac.

It is highly improbable, if not impossible, that anything like rupture of the sac could happen, excepting as the result of injury or great force, in which case the immediate result of the accident would probably call for urgent treatment. Mr. South, however, as late as 1845, described a case in which herniotomy was performed for strangulation, and in which a very complicated sac existed. He says :—

' In carrying my finger around the hernial cavity for
' this purpose (*i.e.* for examination), it suddenly
' passed into an aperture on the outer side, and being
' pushed onwards, entered the large swelling, and
' passing along it nearly as far as the iliac spine could
' be readily felt, and not deeply beneath the skin,
' which was then slit up on my finger, and thereby
' a large mass of healthy omentum exposed, which
' being raised, about four inches of small intestines,
' chocolate-coloured and bright . . . . came into
' view. The mouth of the sac was speedily found,
' and my finger with little difficulty passed into the
' belly. . . . . *I presume* in this case that the hernial
' sac had burst, but how or when the history of the
' case gave no information, and that the protruded
' bowel and large portion of omentum had no proper
' sac, but had merely formed themselves a cavity in
' the cellular tissue.' [1]

---

[1] South's *Translation of Chelius*, vol. ii. p. 76. The italics are mine.

It is unlikely that such a sound observer as Mr. South could have been mistaken ; at the same time, the fact of his only *presuming* upon the existence of this rupture affords an element of uncertainty, which is somewhat increased by the fact that no mention is made of the absence of a *peritoneal lining in the superficial portion of the sac*, which would of course have been the case if the gut and omentum were outside their proper sac.

The reason of the comparative rarity of marked irregularity of the sac, caused by the pressure of strands of fascia, in inguinal hernia is not difficult to find, since the sac passes directly down into the scrotum between layers of fascia, the planes of which are uninterrupted, and all lie in the direction taken by the rupture in its descent, while at the same time there is no tendency to any sudden change in the course of the hernia as occurs in the femoral variety, in which, after the attainment of a certain size, an abrupt bend around the edge of the saphenous opening of necessity happens, unless by chance the hernia forces a way down between the planes of fascia in the thigh—a very rare thing.

An unusual form of irregularity of the sac is described as occurring very occasionally in inguinal hernia, and is due to the coexistence of the direct and oblique form on the same side ; this, however, is rather a *double* sac than a merely *irregular* one. A similar arrangement of great clinical importance is not very uncommonly met with in umbilical hernia, and its existence should never be forgotten whilst operating for strangulated umbilical hernia.

The arrangement referred to is the co-existence with the ordinary large umbilical sac of one or more small distinct secondary pouches, situated in the middle line above or below the main rupture; these little sacs, although perhaps very small, are quite capacious enough to lodge and strangulate a small mass of omentum or knuckle of bowel. On no account, therefore, should the operator in umbilical strangulation omit to pass the finger into the abdomen, in order to ascertain whether any small piece of omentum or gut may at the same time be engaged in one of these little pockets.

Although irregular sacs produced by the pressure of fascial bands are rare in inguinal hernia, the sac often presents deviations in shape which are extremely interesting, more especially in connection with their causes, and also in relation to the frequency with which they lead to strangulation in the sac (see page 26).

These deviations always take the form either of the 'hour-glass' sac or some modification of it. The simple 'hour-glass' form is common in congenital hernia, and in the sac of an infantile hernia I have seen as many as four compartments caused by the projection from the wall of the sac of three imperfect diaphragms, each forming a process of the sac wall itself and quite unconnected with any constricting band of fascia.

The irregularities then in the inguinal sacs are not eccentric and erratic, as in cases where the course of the sac is opposed by irregular fascial planes, but are due to the projection of septa or imperfect diaphragms from the wall itself. In fact, as already stated in considering the formation of valvular flaps and imperfect diaphragms, this variation in form is, I believe, due to the unsuccessful attempts at closure of the funicular process, the sac maintaining its tubular form until the portions between the septa become distended with the hernial contents. Hence such sacs would be expected to exist most commonly in cases of hernia into the funicular process, and this certainly is so, whilst in the ordinary acquired inguinal hernia any marked deviation from the globular or pyriform shape is very rare, with the exception of the very slight depression caused in the sac wall by the transverse layer of fascia already mentioned as lying at the junction of the groin and scrotum. The depression in the sac wall under these circumstances always disappears upon division of this fascia, a fact which probably explains the ease with which some of these herniæ are reduced after the sac has been freely exposed but not actually opened.

LECTURE V

ON SOME DIFFICULTIES AND DANGERS WHICH MAY
ARISE IN ATTEMPTS AT THE REDUCTION OF
HERNIÆ BY MANIPULATION

(Published in the *Lancet*, August 20, 1892)

SYNOPSIS--Introductory remarks—Clinical example (No. XII.), partial laceration of bowel produced by unsuccessful taxis—Comments on this case—Difficulties in the treatment caused by distension of the injured gut in such cases—Author's plan of treatment.

Clinical example (No. XIII.), rupture of the wall of a hydrocele resulting from patient's attempt at the reduction of a co-existing hernia—Comments on this case—Self-inflicted injuries to hernia in attempts at reduction not rare—POSSIBLE DISASTERS WHICH MAY BE CAUSED BY TAXIS—Bruising of the bowel— Laceration of the same—Treatment—Situation of the tear in recent cases usually on the prominent bulging part of the gut, *not at stricture*, as it generally is in advanced or neglected cases—Rupture of adhesion in sac—Rupture of sac itself—Great rarity of this condition—Hæmatocele—Hæmatoma—Reduction *en bloc*.

MANNER IN WHICH ATTEMPTS AT REDUCTION BY MANIPULATION MAY BE SAFELY MADE—Mode of applying taxis—Time which should be employed in its application—Condition of the hernia in its relation to the safe use of taxis—Risk of lacerating gut not always removed even after the sac has been opened—Reference to a case in point—importance of free division of the stricture—Harm more frequently caused by insufficient than by too free division—Difficulty caused in the reduction of hernia by membranous bands and folds after division of the stricture.

IT is, I assume, a matter of common knowledge that many methods of treatment in themselves excellent, and when properly used of much practical utility, may under certain conditions be not only harmful but perhaps disastrous, especially if applied without discretion by any person who does not possess the knowledge or skill requisite for their safe employment. The means available for the reduction of a strangulated hernia without operation form no exception to this rule, and by an unusual chance there are now under my care two patients in the Harris Ward, separated from each other by a single bed only, whose cases illustrate this fact in a remarkable way.

CASE XII.—*Strangulated inguinal hernia; prolonged unsuccessful taxis; herniotomy; gut found partially lacerated; suture of bowel; radical cure; recovery.*

W. F., a man aged 27, was admitted on November 18, 1891. In the previous April he found for the first time that he was ruptured on the right side. The hernia was always reducible, and a truss was continually worn. On the morning of the day before the patient came into the hospital the spring of the truss broke, and he was therefore obliged to work without it. The hernia soon came down, and he was unable to reduce it. Vomiting thereupon set in, and during the night occurred many times. A few hours before his admission an attempt, lasting about a quarter of an hour, was made to reduce the hernia by manipulation, but without success, the patient suffering great pain, especially during the latter part of the trial. The bowels acted regularly up to the morning of the 16th. On admission the man had an anxious look, and was evidently in very acute pain. There was occasional retching, by which a small quantity of sour-smelling gastric contents was

brought up. The tongue was white, but moist; the
pulse 100 and fairly full. In the right side of the
scrotum and running up to the groin was a large,
acutely painful, and tender swelling, very tense, resonant, and entirely without impulse. There was much
pain about the umbilical region, and also some tenderness over the hypogastrium, but no distension. Having
regard to the prolonged taxis to which the hernia had
been already subjected, no further attempt at reduction
by manipulation was made, herniotomy being at once
performed. Upon opening the sac several ounces of
reddish fluid, containing air, escaped. A large knuckle
of greatly distended gut then came into view, deeply
congested, and gently glued to the lower part of the sac
by very recent peritonitis. On freely exposing this gut,
by laying open the sac from end to end, there was found
over its most prominent part (*i.e.* at the anterior aspect
of its lower end) a longitudinal laceration involving the
peritoneum for about an inch; over half its length this
rent passed through the subjacent muscular coat, from
the opening in which the mucous membrane protruded,
and seemed on the point of bursting. The stricture,
which was at the inner ring and extremely tight, was
then very freely divided, with the hope that the tension
in the gut would be thus sufficiently modified to allow
of the wound in its walls being sutured. This, however, was found to be impracticable, as the stitches
immediately cut their way out, the tension not being
sufficiently relieved spontaneously. It was at the same
time clear that any attempt to reduce the tension by
pressing the air contained in the bowel back into the
abdomen would result in the rupture of the protruding
mucous membrane, which I was particularly anxious to
avoid for reasons to be presently discussed. I therefore

selected the healthiest available spot on the bowel, and there punctured it with an exploring trocar, afterwards emptying the knuckle of gut of its contents, which consisted principally of altered blood, through the canula. This having been done, the gut collapsed naturally. The wound in its walls was then brought together with perfect ease and safety by means of five Lembert sutures of silk. A single suture of the same kind having been passed across the small peritoneal wound made by the trocar, the bowel was finally thoroughly cleansed and returned into the abdomen. The neck of the sac was then isolated, ligatured, and invaginated after it had been severed from the scrotal portion; the pillars of the ring were closed by kangaroo tendon sutures, a drainage-tube inserted, and the superficial wound united with silkworm gut. On the 20th four loose motions containing altered blood were passed. The bowels acted daily after this, and by December 18 the patient was walking about the ward apparently in perfect health. In fact, he was practically well by December 4, but being very apprehensive about himself, he was allowed to remain in bed longer than was really necessary.

In addition to the points specially relating to the main subject of the present lecture, this case affords a good illustration of the successful treatment of cases of the kind by a method to which I wish to call particular attention. In all instances of strangulated hernia in which the laceration of the bowel does not involve the whole thickness of its walls, so that the intestinal canal is not opened up, the gut as the result of the strangulation is of course greatly distended. In attempting to bring together the edges of the wound under these circumstances, difficulty generally arises in consequence of the sutures cutting their way so readily through the peritoneum, that sufficient traction cannot be made upon them to effect the closure of the laceration with any safety. This tendency on the part of the stitches to cut their way out is due partly to the unyielding tension of the bowel and partly to the softening of its coats from pathological changes. Not only, therefore, is very little pressure required under such circumstances to increase the rent in the peritoneum, but when the tear involves the whole thickness of the muscular coat the mucous membrane bulges through the wound, and extremely slight force only is necessary to cause the protruding portion to give way, thus converting an incomplete laceration into a wound opening up the intestinal canal.

In cases of this sort it is sometimes also found that even after free division of the stricture no sufficient alteration of the tension occurs until digital pressure is used, which may be quite sufficient to burst the stretched and protruding mucous membrane —an accident which it is most important to avoid if possible, not only because it must of necessity allow the bowel contents to escape over the wound, but also because the protruding mucous membrane is usually so thin and injured that if it is damaged further by laceration, incision, or puncture, ulceration, and perhaps sloughing, is prone to occur later on, and give rise to trouble which may possibly be serious. In order to diminish the tension of the bowel and to allow of its shrinking of its own accord without inflicting any further injury upon the protruding mucous membrane, I adopt and confidently recommend the following plan, which, it will be noticed, was used in this case.

The stricture having been freely divided, and the distended knuckle of gut carefully cleansed, an exploring trocar is passed into the bowel through the portion of its walls which appears most healthy (*i.e.* at a point which is certain to be somewhat remote from the laceration). The contents of the gut, consisting generally for the most part of altered blood, pass out easily through the canula into a suitable receptacle, and are thus conducted quite away from the peritoneum, which therefore remains entirely unsoiled, as the muscular coat at the point of puncture clings so closely to the canula that no leakage takes place. On the other hand, if the puncture be made in the protruding mucous membrane, leakage is certain, and soiling of the whole rent occurs, in addition to the injury inflicted by the puncture on the damaged structures. The collapse of the gut which follows spontaneously makes the approximation of the peritoneal surface over the laceration perfectly easy and safe, no matter how softened the intestinal walls may be. The gut may then be returned without hesitation, after it has been finally cleansed and a single Lembert suture placed across the little nick in the peritoneum at the seat of the puncture. This last detail is perhaps hardly necessary, but it is a source of additional safety during the return of the bowel, and adds nothing materially to the length of the operation. The method here described of relieving tension in damaged or partially lacerated bowel by puncture through the healthier parts rather than through those which are most injured has, so far as I know, not been previously insisted upon, but I am sure it is worthy of adoption.

CASE XIII.—*Inguinal hernia with large hydrocele; attempted reduction by patient; rupture of wall of hydrocele, which was thus converted into a hæmatocele.*

W. B., a labourer, aged 53, was admitted into Harris Ward on November 21, 1891, with the following history. Three years previously he commenced wearing a truss, as he was told he had a rupture; and about eighteen months later he noticed some swelling of the scrotum which, upon applying to the Truss Society, he was told was not due to a hernia. On the day before he came to the hospital he felt some pain about the groin, and noticed more swelling there than he had previously seen. Thereupon, thinking it was the hernia which had come down, he attempted to put it back by seizing the whole scrotal tumour in his hands and using considerable force. No

G

immediate alteration was produced in the swelling, but acute pain followed directly, and in a few hours the whole scrotum became almost black. The pain gradually decreased, but much tenderness followed, on account of which he applied for treatment. On admission the patient was a big, strong-looking man. The left side of the scrotum was much discoloured and distended by an oval swelling, which was very tense, tender, and non-translucent. Its upper end was quite free from the abdominal ring, through which a hernia came down when the patient coughed. The bruising slowly disappeared, and the swelling decreased a little in size after rest in bed. On December 17 I laid open the scrotal tumour, letting out a quantity of hydrocele fluid mixed with altered blood. On examining the walls of the cyst there was found at the upper and inner part a large rent, admitting the ends of three fingers, which passed into a considerable space beneath the scrotal integuments. The parts about this were matted together by recent inflammatory material, and the walls of the hæmatocele were covered with new lymph. The hernia was not exposed in the operation, as it was thought wiser, considering the inflamed condition of the tissues, to postpone its radical cure, which the patient was anxious to have performed, until a subsequent date. The patient naturally made an uninterrupted recovery.

The main point of individual interest in this case is the fact that the injury was self-inflicted. Although the lesion produced here by the patient's manipulation was not a serious one, it is so sometimes. I have myself had an opportunity of seeing an instance in which taxis applied by the patient

resulted in a small rent in the gut through which
some fæcal material had escaped into the sac. It is
true that the gut at the seat of injury had obviously
been adherent, the laceration having resulted from
the tearing away of the adhesion by the force
applied ; the fact, however, is significant that laceration of the bowel itself was produced by no more
violent manipulation than that applied by a patient
perfectly accustomed to reduce his rupture.

The main disasters which may immediately
follow upon attempts at the reduction of hernia
by taxis may be summarised as follows :—(1) Bruising of the bowel; (2) rupture of the bowel walls,
complete and incomplete; (3) laceration of adhesions ; (4) rupture of the sac ; (5) hæmatocele (in
cases complicated with hydrocele); (6) hæmatoma
of the scrotum and surrounding parts; (7) reduction
of the whole hernia with the sac (reduction *en bloc*).

1. *Bruising of the bowel.*—Some bruising, as
shown by sub-peritoneal extravasation, of large or
small extent, in the walls of the herniated bowel will
be found in the majority of cases which have been
submitted to taxis, unless extreme gentleness has been
used. The extent and severity of the injury will
naturally depend, for the most part, upon the amount
and direction of the force applied and to a considerable degree upon the condition of the gut, which
bruises more readily when greatly distended, especially

in neglected cases which have been allowed to continue for a long period without treatment. It is interesting to note that bruising of the bowel, if at all extensive, although no apparent breach of surface on the peritoneal aspect may exist, is almost invariably associated with bleeding into the intestinal canal, a fact conclusively demonstrated in many cases by the appearance of altered blood in the first motion passed after the relief of the stricture. It may, in fact, be accepted without reserve that attempts at reduction by manipulation produce some bruising of the bowel in the great majority of cases of strangulated hernia. At the same time it may be fairly admitted that, as a rule, unless great carelessness has been used, no permanent harm results. It is, nevertheless, necessary to insist on the occurrence of injury from this cause, in order to give weight to the fact that attempts at reduction by manipulation are liable at times to produce damage. Taxis, therefore, should not be regarded, as it seems to be by some people, as a plan of treatment which, if it fails to reduce the rupture, at least can do no harm. This last remark must not be taken to imply any objection to the proper practice of this method as such, but rather as a warning against its use carelessly and without due regard to possible evils which may, under certain conditions, result.

2. *Laceration of the bowel.*—This may, of course,

involve the whole thickness of the intestinal wall
or only one or more of its coats; the former is
naturally the most serious, since it allows of the
escape of fæces into the sac. The latter condition
may vary in degree from an almost imperceptible
crack in the peritoneum to a laceration in the
peritoneal and muscular coats inches in length, in
which case the mucous coat protrudes, hernia-like,
through the opening in the muscular wall. Whether
the laceration is partial or complete the treatment is
identical. The edges of the wound must be brought
together with Lembert sutures, care being taken
that the end stitches—if the lesion is of any extent—
are placed a short distance beyond the extremities
of the rent. When the gut is too tense to allow of
approximation of the peritoneal surfaces it should,
if necessary, be emptied in the way I adopted in the
first of the cases which form the basis of this lecture.
However small the crack in the peritoneum is, even
if it be hardly perceptible, a single suture should be
passed across it. If the condition of the patient in
cases of partial laceration is so desperate that the
delay entailed by the suturing process is not justifiable, the gut should be cleansed and returned into
the abdomen, a drainage-tube of large calibre, *without
lateral perforations*, being placed in the canal in the
manner described in Lecture VI.

A point of great interest, to which sufficient

attention has not, I think, of late been paid, arises here with reference to the situation at which laceration of the bowel from injury occurs in these cases. There appears to be an idea, which is traditional and supported apparently by the teaching of the schools now, that the tear produced by injury in the gut of a strangulated hernia takes place *at the seat of stricture* in consequence of the way in which the sharp edges of the constricting tissues, as it were, cut into the distended bowel when pressure is exerted upon it.

Now I have seen several cases myself in which a rent in the gut was undoubtedly produced by taxis, and in two of these the lesion *was not at the point of constriction but on the prominent bulging and most distended portion of the bowel.* Both of these cases were recent, and in each the rent was in the long axis of the gut. The same result followed in some experiments made by me on the cadaver, artificial strangulation in two cases of old herniæ having been produced by inflating the bowel from the abdomen and forming a stricture by ligaturing the neck of the sac together with its contents. The hernia was in one instance then violently crushed and in the other struck sharply with a stick : in both the laceration occurred *on the prominent part of the strangulated knuckle and in the long axis of the gut.* On consideration, this result, so far as the situation of the injury is concerned, is, I think, precisely what should be expected in recent cases, for in such the rent begins in the peritoneum, which under pressure naturally gives way at the weakest point (that is to say, where it is most thinned and stretched by distension). A sudden blow, therefore, or prolonged hard pressure would, as a matter of course, lacerate the peritoneum in the part most stretched and thin— *i.e.* over the end of the distended knuckle rather than at the seat of constriction, where not only is it unstretched, but where it is actually supported by the surrounding parts.

In cases far advanced and neglected the state of affairs is altogether different, because in them the gut at the seat of stricture is indented by the edge of the constricting tissues, perhaps partially eaten through by ulceration from within, or possibly gangrenous and on the point of giving way. Then the weakest point is at the strictured part, and very little force may be necessary to complete the perforation which has already commenced. It is, I presume, in connection with cases of this latter kind that the traditional teaching has been fostered, for in recent cases of strangulation it certainly does not apply.

3. *Rupture of adhesions in the sac.*—The tearing of recent adhesions during attempts at reduction by manipulation need not have any bad result, but

free hæmorrhage into the sac may thus be produced so
that it may be completely filled with blood, although
no serious lesion may be apparent. In old irreducible
herniæ, in which band-like adhesions sometimes exist
between the bowel and the sac wall or between
different parts of the bowel itself, no harmful results
need follow if the adhesions themselves give way, but
if, as may happen, an adhesion is torn away with
some of the intestinal peritoneum, a partial laceration of the gut results, which is, if at all extensive,
a serious accident, especially if the case has been
neglected and operation long deferred.

4. *Rupture of the sac.*—This, although a recognised injury and classed as one of the modifications
of the reduction *en masse*, must be a very rare
sequence of taxis, as the amount of force required to
tear the sac is very great and would hardly be inten-
tionally applied. Bruising of the sac is, however,
common, and I have seen a portion of its wall torn
away with an omental adhesion; this, however, is
not a rupture in the sense under discussion—*i.e.* a
splitting of the sac wall from sudden or gradual
pressure. In the *post-mortem* room I have not been
able in artificially strangulated herniæ to rupture the
sac without also bursting the gut. Rupture of the
sac is therefore probably too rare to be other than a
curiosity; indeed, Sir Astley Cooper, after his large
experience, says that it ' scarcely ever ' occurs from.

any cause. Formerly spontaneous rupture of the sac was also a recognised condition, but actual evidence of its existence seems wanting, or at all events is not convincing.

5. *Hæmatocele.*—A good example of this accident is afforded by Case XII., described in this lecture. After what I have said concerning the difficulty in causing rupture of a hernial sac, it may at first sight seem strange that the sac of a hydrocele should give way so easily. There is nothing inconsistent, however, in this matter, for it must be borne in mind that the sacs of hydrocele sometimes undergo pathological changes which result in softening and thinning, so that they become weak in parts.

6. *Hæmatoma from rupture of a large vein or veins outside the sac.*—Enormous blood extravasations may be thus produced, and occasionally upon the application of comparatively slight force only, especially in elderly people. I once had an opportunity of seeing a large blood swelling involving the scrotum and groin, which were nearly black from discolouration, said to have followed upon nothing more violent than the manipulation necessary for the adjustment of a truss to an easily reducible hernia in a patient nearly eighty years old.

7. *Reduction " en bloc "*—i.e. *the reduction of the sac, together with its contents, the strangulation being therefore unrelieved.*—The only point to which I need

here call attention in this respect is the singularly small amount of force which sometimes seems necessary to produce this accident. I have personally seen only one case, and that was not in my own practice. The hernial tumour seemed to disappear almost the moment the hand was laid upon it, and certainly before there was time for the application of any methodical violence. So much was this the fact that I cannot help feeling that there must be in such a case some kind of spontaneous action from above which contributes to the reduction. Is it possible that extreme irregular spasmodic attempts at peristalsis may act in this way?

It is only fitting that such an ominous list of casualties, which are not only possible, but actually occur in practice as the result of the use of taxis, should be followed by some indication as to when and how the reduction of a hernia by manipulation may be attempted without risk. The safety of this plan of treatment is dependent on three conditions. (1) The manner in which the taxis is applied; (2) the period during which the manipulation is persisted in; and (3) the state of the hernia.

1. *The mode of applying taxis.*—This may appear such a purely elementary point as to render its consideration hardly justifiable outside the pages of a student's text-book. It is nevertheless true that practitioners, otherwise intelligent and trustworthy,

do at times manipulate a hernia in the manner best calculated to cause injury to the contents of the sac, whilst it affords the least possible chance of effecting reduction. I do not propose to occupy time now with a minute description of the method by which the taxis may be applied safely and with a fair prospect of success, as it can be usefully learnt only from practical demonstration at the bed-side, but some of the details of the process are so important and essential that they require a passing notice. The details referred to are as follows : (*a*) All manipulation should be conducted only with thoroughly warm hands ; (*β*) the neck of the hernia should be firmly supported by one hand whilst the other manipulates the body of the tumour. (*γ*) In using the fingers all pressure from the finger-ends should be made by *the front of the digital pad and never by the actual tips.* (*δ*) The pressure necessary in the manipulations should be gentle, firm, and regular, not forcible, unsteady, and spasmodic.

The necessity for warm hands, for the support afforded to the neck of the hernia, and for the avoidance of the use of the actual finger-tips, is, I cannot help feeling, not so universally realised as it certainly should be, for I have more than once seen attempts made at the reduction of a rupture by grasping the body of the tumour with hands almost blue with cold, the neck of the hernia being left entirely unsupported; then with a punching and rolling movement, during which the finger-tips have been deeply pressed into the parts, the force has been gradually increased until further persistence in the attempt has been rendered impracticable by the protests of the patient. Where injury is possible, it is from some such faulty plan as this that it is most likely to result. The cold hands excite every resistance in the way of muscular action; the want of support to the neck of the hernia makes its reduction very unlikely by allowing the gut to bulge over the margins of the constricting ring; and, beyond this, in neglected or long-standing cases, when the bowel has commenced to ulcerate from within, the pressure of the sharp edges of the stricture acts at a great advantage in further injuring and perhaps bursting the thinned and weakened intestinal walls. Finally, the sharply indenting finger-tips are admirably adapted for causing an unnecessary amount of bruising and possibly laceration of the gut.

2. *The time which should be occupied in taxis.*— Judging from my own experience, and from what I have seen in the practice of others, five minutes should be taken as the outside limit during which manipulation of a hernia in cases of apparent strangulation may be with safety persisted in, no matter how gently it is applied. In unstrangulated cases the same time should always be considered as suf-

ficient, for, although no actual harm need result, if the time be extended it may very easily be brought about; moreover, if success is not attained by the end of five minutes it is very unlikely to result at all, and further attempts are practically useless.

3. *The condition of the hernia.*—When properly applied and with the precautions just mentioned, taxis may be used with safety—(*a*) in all cases in which the true hernial impulse is present, provided always that there is neither any marked tenderness nor inflammation in the sac or its contents, when its employment would, of course, be entirely negatived; (*b*) in very recent cases of strangulation where the tension is not extreme. This latter is a recognised principle and is, therefore, worthy of respect; but I very much doubt whether it is possible, excepting perhaps in infants, to reduce by manipulation any rupture in which the hernial impulse is not present. For myself, at least, I must admit that I have never been able to return with any reasonable application of force a hernia in which I could detect no impulse. This impulse, it is true, may have sometimes been slight, but it was present invariably in the cases where reduction was possible, although it must be admitted that I could not always demonstrate it in the hospital patients to my house surgeons. A large distended hernia universally resonant should be

treated with more than usual gentleness, for in such cases the bowel is far more liable to injury than in any other kind, especially if adhesions happen to exist in the sac. Hernial tumours, dull on percussion, with omental or fluid contents, may be manipulated with greater freedom without much risk of damage being done, but in these reduction is entirely out of the question in the absence of impulse; the utility, therefore, of persistence in the attempt at all under these circumstances is not plain.

Every case of apparently strangulated hernia must necessarily be treated upon its individual merits; but, for my own part, I am sure that, as a general principle, it is better in herniæ which are obviously strangulated and entirely without impulse to perform herniotomy at once rather than make attempts at reduction by manipulation, because I have no doubt whatever that early herniotomy in fairly competent hands is infinitely less hazardous than an unwise persistence in fruitless attempts at reduction by taxis. If due regard be paid to the patient's welfare one thing at least is certain—viz. that a strangulated hernia which has already been subjected to taxis should be operated upon at once, and no further manipulation used until after the tumour has been explored and the stricture freely divided.

It must not be imagined that all risk of lacerating

the bowel during attempts at its reduction necessarily
ends after the sac has been laid open in herniotomy,
or, indeed, in *every* case after the stricture has been
divided, for, although to the best of my knowledge
the accident has not occurred under these circumstances in my practice, I have been present at an
operation in which a surgeon of experience certainly
did produce a laceration in the peritoneal coat of the
bowel whilst attempting to reduce it after the division
of the stricture, which had obviously given rise to the
strangulation. This difficulty, sometimes experienced
in reducing the hernia after the stricture has been
cut, is undoubtedly as often as not due to the division
being not sufficiently free, the little nick so commonly recommended being too slight for securing
the necessary relaxation of the constricting band. I
am sure, from my own observation, that harm is
more often likely to arise from too slight a division
of the stricture than from one which is too free.
Free division of the parts about the neck of the
hernia as a rule entirely obviates any chance of
injury to the gut, whilst the possible anatomical
dangers entailed in this free incision have been, I
have no hesitation in saying, unduly exaggerated.

Although I make a practice of dividing the
stricture freely, I have never had the slightest cause
to regret it, and certainly have never seen any
hæmorrhage which has given the least anxiety

afterwards. The only case in which I have had any trouble whatever on account of bleeding after herniotomy was a strangulated umbilical hernia, in which alarming hæmorrhage took place into the abdominal cavity from a torn omental vein. This vessel was almost certainly burst by the force which was necessary for the return of the hernia through a ring which had been insufficiently divided; had the division been altogether more free the hernia could have been reduced without any force, and the vessel would, I believe, have undoubtedly remained intact.

I must here again refer to a point which is especially interesting in connection with a further difficulty which occasionally arises in the reduction of a hernia, even after the stricture has been freely divided. At first sight it is singular that any difficulty of this kind should occur at all, still it is quite certain that it is sometimes met with. For instance, in a case of inguinal hernia under my own care I was unable, after repeated division of the stricture, to reduce the intestine, although on passing the finger, as is my habit, through the canal into the abdominal cavity, I could feel nothing in any way constricting the bowel. The only noticeable thing to be felt was a loose membranous fold which, springing from the outer wall of the canal, lay quite flaccid upon the gut, and allowed my finger to pass

by it with perfect ease. Whilst I was attempting to
return the bowel, the end of one finger being placed
on it just below this fold, I noticed that as the
gut was pushed against the flaccid flap it seemed
to grip the bowel after the manner of a sling. I
therefore divided the fold, and returned the hernia
without the least trouble. Here, then, the obstacle
to reduction was clearly this loose sling-like fold.
The existence of membranous flaps like this, and the
manner in which they sometimes resist the return of
the gut in operations for strangulated hernia, have
not of late received sufficient attention. Bands
and flaps of this kind, which are not very rare,
should invariably be divided whether they seem to
compress the bowel or not, for if they do not actually prevent reduction it will be much more easily
effected after their division. The history and mode
of formation of these folds have been fully considered
in Lecture IV.

LECTURE VI

ON THE MANAGEMENT OF DAMAGED BOWEL IN STRANGULATED HERNIA

(Published in the *Lancet*, October 18, 1890)

SYNOPSIS—Introductory remarks—Clinical example (No. XIV.), Lacerated bowel in a case of strangulatedfe moral hernia—Clinical
example (No. XV.), abrasion and extensive bruising of the bowel
in a case of femoral hernia—Clinical example (No. XVI.), strangulated femoral hernia with apparently gangrenous gut—Remarks on
the treatment of these cases—Rapid repair after suture in laceration dependent on three conditions—(1) healthy state of intestinal
coats—(2) Accuracy of suture—(3) Careful cleansing of parts before
reduction—Best form of suture—Reference to case showing the
strong tendency to healing existing in these cases—Contents of
bowel not necessarily septic—Remarkable nature of recovery in
Case XVI. without fæcal fistula—Explanation of this.

Treatment of gangrenous or very badly-injured gut—Objection to immediate formation of artificial anus—Author's treatment by returning
gut into belly and leaving large drainage-tube in canal—Resection of
damaged gut in strangulated hernia.

APPENDIX—Case XV. regarded as one of incomplete laceration from
taxis—Reasons for not treating the injured gut by tapping and
suture, as recommended in Lecture V.—Further reference to the
treatment by leaving large drainage-tube in canal—Objections raised
to its use not valid—Necessity for tube being unperforated laterally
—Reasons for this—Object of the tube.

Question of desirability of dividing the stricture when gangrenous gut
is left unreduced—Author's views—Reasons for the same—Further
reference to resection of gut in these cases.

IT is hardly necessary to say that the prospect of the
recovery of patients after herniotomy for strangulation of the gut depends, to a large extent, upon the

condition of the contents of the sac, especially the bowel, at the time of operation. Amongst the cases upon which I have operated in the hospital during the past twelve months, there are three which have such an important bearing upon this point that they merit careful consideration, because they afford excellent illustrations of some of the complications which may at any time call for treatment. Each of these cases possesses some special point of interest, and the last two are more particularly worthy of note in connection with the present inclination to adopt severe surgical measures, such as resection of portions of gut, when the bowel appears to have been damaged by violence, ulceration, or gangrene, beyond the possibility of repair by natural means.

Case XIV.—*Strangulated inguinal hernia; herniotomy; laceration in bowel; reduction after suture; rapid recovery.*

Jas. T——, aged 50, admitted under my care on February 19, 1889, had been operated upon for strangulated hernia in the right groin by Mr. Henry Lee in 1864, when he recovered quickly, and left the hospital provided with a truss, which, however, he only occasionally used. Since that time the rupture had usually been down, being always easily reducible till the morning of his coming under my care, when suddenly, during an attack of coughing, the hernia became much larger, very painful, and irreducible. Vomiting followed almost immediately. On admission the man was pale and anxious-looking. There was a large irreducible right scrotal hernia, very

tense and tender. Slight impulse was felt in the tumour
when the patient coughed; there was occasional vomiting,
and flatus only was passed per anum. Under treatment
the vomiting ceased. In twenty-four hours all impulse
had disappeared and no flatus was passing. Herniotomy was therefore performed on February 20, through
the cicatrix of the former operation, which was evident
on the front of the tumour. The sac contained a large
quantity of ill-smelling bloody fluid and a knuckle of
small intestine. Upon dividing the stricture, which was
very tight, and raising the gut, a quantity of fæcal
matter was found in the back of the sac. On the
posterior aspect of the bowel was a slightly lacerated
wound about an inch and a half long, involving all the
intestinal coats, the opening in the mucous membrane
being about half the length of that in the peritoneum.
The parts having been thoroughly cleansed with corrosive sublimate solution (1 in 1,000) the wound in the
mucous membrane was brought together by means of
four silk stitches, and the peritoneum closed over it with
seven Lembert sutures (silk). After free irrigation the
sutured knuckle of gut was returned into the abdomen
and the ring closed. No drainage was used. The herniotomy wound healed in four days. The bowels acted
spontaneously within thirty-six hours of the operation,
and continued to do so with daily regularity. The
patient was detained in the hospital till March 27 by an
attack of bronchitis, which, he stated, had commenced
before his admission.

CASE XV.—*Strangulated femoral hernia; herniotomy; partial laceration from taxis; gut reduced, although
too much damaged for suture; rapid closure of resulting
fœcal fistula.*

Sarah P——, aged 62, was admitted under my
care on July 6, 1889. She had been ruptured for ten
years on the right side, but had only occasionally worn
a truss, as it did not keep the rupture up. Three days
before admission the hernia suddenly became painful,
and greatly increased in size during an attack of
coughing. Frequent vomiting followed, and all attempts
at reduction having failed, she was sent to the hospital.
On admission there was extreme collapse ; the expression
was 'abdominal' and the pulse feeble ; the vomiting
was continual, but not stercoraceous. In the right groin
extending upwards over Poupart's ligament was a
large, tender, irreducible tumour, without impulse.
Herniotomy was performed at once. The sac contained
some very dark fluid, a considerable mass of omentum,
and a deeply congested knuckle of small intestine. On
the front of the gut was a breach of surface involving
the peritoneum and part of the thickness of the muscular
coat for about half an inch. The stricture was freely
divided, the omentum ligatured and removed. It was
then found that the slight pressure exerted upon the
bowel in these proceedings had caused a considerable
extension of the laceration in the peritoneal coat, which
the slightest manipulation still further increased. On
attempting to bring this wound together by means of
Lembert's suture, the stitches immediately tore through
the softened tissues unless they were placed at such a
distance from the edges of the rent that a dangerous
kink in the bowel resulted, if the peritoneal surfaces

were properly approximated, which was very difficult.
After careful cleansing the gut was therefore returned,
without any attempt at bringing the wound together,
just inside the ring (although it was perfectly plain to
my mind that it must give way), and a large drainage-
tube placed in contact with it. The herniotomy wound
was then closed with horsehair sutures, and dressed
with the double cyanide gauze. The wound healed by
first intention, excepting of course around the tube.
The bowels acted spontaneously on the seventh day
after the operation. Three days later healthy fæcal
matter passed along the tube, causing no irritation or
rise of temperature. The bowels subsequently acted
regularly every day; by August 18 the fæcal fistula,
which was dressed antiseptically throughout the case,
had completely closed, and the patient left the hospital
wearing a truss. She has since been seen (six months
after the operation), perfectly well in all respects,
and without any apparent tendency to a descent of the
rupture.

CASE XVI.—*Strangulated femoral hernia; herniotomy; reduction of obviously gangrenous gut; recovery
without fæcal fistula.*

Eliza J——, aged 55, was admitted under my care
on December 22, 1889. She had always been in good
health, and was not aware of having suffered from
rupture. Four days before admission, whilst at her ordinary work, she felt great pain in the right groin, and
noticed for the first time a hard tender swelling in that
locality. The pain increased, being especially severe
about the umbilicus. Vomiting soon came on. Free
taxis having failed to reduce the tumour, she was sent
to St. George's Hospital. On admission the patient was

greatly collapsed.   There was almost continual vomiting.
The pulse was small and quick, the tongue dry, and the
urine albuminous.   In the right femoral region, passing
on to Poupart's ligament, was a very tense, tender, and
irreducible swelling, as large as a billiard ball, without
impulse on coughing.   Herniotomy was performed immediately.   The whole of the front of the sac was stinking and gangrenous ; it contained no fluid, being
entirely occupied by a knuckle of small intestine, which
over its central part was piebald in aspect, quite lustreless, and, so far as could be judged, obviously gangrenous.   The sloughing sac was cut away, the stricture
very freely divided, and the gut, although clearly
damaged beyond chance of recovery, returned as gently
as possible just inside the ring, after having been freely
irrigated with corrosive sublimate solution (1 in 1,000).
A large drainage-tube was passed through the ring
and left as near the returned bowel as possible.   The
herniotomy wound having been closed with horsehair
sutures, a ' double cyanide ' dressing was, as usual, carefully applied.   When removed from the operating table
the patient seemed nearly moribund, and I had little
hope of her recovery.   Contrary, however, to our ex-
pectations, she rapidly rallied.   The wound healed by
first intention excepting around the tube, through which
for a week ill-smelling discharge of grumous appearance
came in small quantities.   No fæcal material was at any
time detected.   The bowels acted spontaneously on the
fifth day after the operation, the motion being almost
black in colour, but not unhealthy in odour.   The tube
and stitches were removed on the eighth day, the discharge by that time having become serous in character
and quite inodorous.   Some bronchitis from which she
had long suffered kept the patient in bed for a fortnight

longer, but by January 16, 1890, the parts were perfectly
sound, and a truss was fitted. She was seen again two
months later in perfect health, no complication of any
kind having arisen.

In this case, and also in the previous one, the
question of resection of the damaged portion of the
bowel was naturally considered. It was, however,
evident that neither of the patients was in a condition
to stand the shock of a prolonged operation, even if
it had been considered that the state of the parts
locally rendered any such proceeding justifiable.

Taking the three cases which I have described in
the order in which they came under treatment, the
main points of interest presented by them are the following:—Case XIV. is an excellent and simple instance
of the good result obtainable by the accurate suturing of an extensive wound involving the whole of the
coats of healthy bowel, and subsequently returning
the gut into the abdomen, after careful cleansing
from the fæcal material extravasated about it. As
will have been seen, the rapidity with which union
occurred in this case, although not exceptional, is
remarkable, for the bowels acted spontaneously about
thirty hours from the time of the operation without
any harmful result. The rapidity of repair in all
such cases depends, for the most part, upon three
conditions: (1) a healthy state of the injured intestinal coats, (2) accuracy of suture, and (3) careful

antiseptic cleansing from all contaminating material. The first of these appears to be essential for rapid and sound healing. As to the form of suture, in my experience the Czerny-Lembert is the best. The actual form of suture, however, so long as an accurate apposition of parts is obtained, is probably of less importance than the careful cleansing of the gut before it is returned, and nothing seems better for this purpose than a solution of corrosive sublimate (1 in 1,000). At the same time it is only fair to admit that repair equally rapid is quite possible, although, of course, less probable, without any peculiar kind of suture or special method of cleansing. Some years ago Mr. Holmes,[1] in a case similar to the one I am now discussing, brought the edges of the wound together with sutures passed through and through the whole thickness of the bowel walls, and returned the gut after merely wiping it with a wet sponge ; recovery followed almost as quickly as in this case of mine. The instance quoted is worthy of more than a passing notice, not only because it shows how strong is the tendency to safe repair in some of these patients, when the gut is healthy, but also because it appears to confirm, to some extent, the impression that the contents of the bowel are not in themselves necessarily septic, at all events when coming from the small intestine.

[1] *St. George's Hospital Reports*, vol. ii. p. 825.

Cases XV. and XVI. are extremely interesting from another point of view. Case XV. is an excellent example of the production of a fæcal fistula by the giving way of the injured bowel, and its subsequent spontaneous closure; a very ordinary occurrence in neglected strangulated herniæ in the groin, but one which does not always receive enough attention, in relation to the treatment of damaged gut in such cases. Case XVI. is a remarkable example of unexpected success in surgery. I can state with absolute confidence that no surgeon would have had the least hesitation in condemning the implicated gut as gangrenous beyond hope, for, in addition to the other symptoms, there were present over the central part, in the words of Sir James Paget,[1] ' colours ' about which there can be as little doubt for signs ' of gangrene—white, grey, and green—all dull and ' lustreless.' 'Intestine with these marks,' continues the same authority a little later, ' even though they ' be small, must not be returned.' Admirable in all respects as this dictum may have been at the period when it was so graphically laid down, it seems to me that it may now be somewhat modified. Certainly in this case nothing could have been more perfect than the recovery of the patient, in spite of the reduction of the manifestly gangrenous gut. In another instance in which a knuckle of bowel having

---

[1] *Clinical Lectures and Essays*, 2nd edition, p. 144.

a small area of undoubted gangrene in its walls was returned, after careful washing, recovery followed upon the formation of a fæcal fistula, which discharged fæces for forty-eight hours only, and was soundly healed in nine days. The recovery of Eliza J——— (Case XVI.) without a fæcal fistula can only be explained in one of two ways—(1) that the symptoms of gangrene, perfect as they seemed to be, were deceptive, which is most improbable, or (2) that by reason of the thorough cleansing of the affected bowel the gangrenous portion was enabled to gradually disintegrate and come away through the large tube used, without the formation of any gross interruption of continuity in the intestinal wall. The treatment adopted in Cases XV. and XVI. is worthy of attention, as it differs in some respects from that which is commonly recommended under the circumstances met with.[1] As already mentioned, I regarded the bowel in both instances as damaged beyond the possibility of recovery. Setting aside any attempt at resection of the injured parts, which was altogether out of the question, two courses in each case were obviously open, one of these being the formation of an artificial anus after free division of the stricture, the gut of course being left unreduced; the other being the return of the gut into the

---

[1] See, for example, Mr. John Wood in *Ashhurst's Encyclopædia of Surgery*, vol. v. p. 1135.

abdomen, with a view to the immediate production of
a fæcal fistula.   In deciding as to which is the better
of these courses to adopt, in any given case, it is
especially important to bear in mind that an artificial
anus shows, as a rule, little tendency to spontaneous
closure in consequence of the obstacle afforded by
the projecting spur, produced by the abrupt bend in
the bowel, which is unavoidable when the gut remains
engaged in the neck of the sac.   The cure, therefore,
of an artificial anus can, as a rule, only be effected
by surgical measures, which may be severe and are
notoriously uncertain.   On the other hand, if the
intestine be returned into the abdomen, not only is
there some chance of its complete recovery, if such
is possible, as happened in the case of Eliza J——,
but, upon its walls giving way, a fæcal fistula follows
which will almost certainly close of its own accord
if means be provided for the immediate and free
escape of the extravasated fæces from the abdominal
cavity.   This object is easily obtainable by the adjustment of a drainage-tube of large calibre (not less
than half an inch in diameter), so that its inner end
may lie just inside the abdomen—*i.e.* almost in
contact with the returned knuckle of bowel, which
remains after reduction close to the deep orifice of
the canal through which it has passed, and very
soon becomes adherent to the surrounding parts.
It is needless to say that the result following the

deliberate production of a fæcal fistula, if it can be with safety brought about, is far superior to that which ensues upon the formation of an artificial anus, provided of course that the cases selected are suitable ones.

Should gangrene involve a considerable length of the gut, or if separation has actually commenced, the bowel having given way in consequence, the formation of an artificial anus is clearly the only safe treatment, as cases of this kind are almost always eminently unsuited for resection. At the same time, if the gangrenous area happens to be small, I do not hesitate to return the bowel and use the large tube in the way I have described, even if a small perforation exists, so long as the opening can by temporary sutures or any other means be kept closed for a time judged to be sufficient (probably six or eight hours), to allow the damaged gut to contract adhesion to the parts around.

The main points, then, of the plan which I advocate for the treatment of damaged bowel of the kind now under discussion, two examples of which are afforded by Cases XV. and XVI., are the following: (1) Free division of the stricture in order that (2) the intestine may be returned just inside the abdomen, with as little pressure as possible, after thorough antiseptic cleansing; (3) the careful adjustment of a drainage-tube, having a diameter of about half an inch, in such a way that its inner extremity may lie almost in contact with the reduced bowel.

This is a detail of much importance, since free and immediate escape of fæces from the abdomen, if the bowel gives way, is not only conducive to the rapid healing of the fistula, but is necessary at first for the avoidance of risk of general peritonitis from the passage of retained fæces into the peritoneal cavity, and later for the prevention of fæcal abscess with its possible complications of peritonitis, cellulitis, or intractable sinuses. The tube should remain undisturbed certainly until the bowels have acted for the first time after the operation, and it is safer that it should not be removed before the second motion has passed, for in the less serious cases of damage to the gut followed by fæcal fistula no escape of fæces as a rule takes place until the bowels have acted freely. If a second action occurs without a fæcal discharge from the wound, it is fair to infer that the intestine has recovered itself, and the tube may be safely dispensed with. The continual use of carefully applied antiseptic dressings throughout the case from the time of operation until the final closure of the fistula, if such has formed, is essential. This is especially important during the escape of fæces, in order that these, if not septic when leaving the gut, may be prevented from becoming so whilst lying in the wound.

One word in conclusion with reference to the resection of damaged bowel in strangulated hernia. Personally I have never met with a case in which the operation was worthy of serious thought, and I fancy the experience of most other operators must be similar. The truth of the matter being that in almost every instance where the damage, whether it be from injury or disease, is too extensive for treatment upon some such lines as I have explained in this lecture, the patient is in too critical a state constitutionally to bear the necessary prolongation of the operation entailed by the excision of the affected parts, even though the unavoidable extension of time be reduced to a minimum by the use of contrivances like Senn's decalcified bone plates.

I do not, of course, mean to say that the coexistence of local and constitutional conditions rendering the operation justifiable is impossible, although I believe it must be singularly rare. Should such a case fall into my hands, the operation would receive the consideration it justly deserves, as the result which follows directly upon the successful resection of the affected part of the gut is as brilliant as anything in surgery can well be. In attempting to arrive at a sound decision upon a treatment which, when successful, gives such a tempting result, it must never for a moment be forgotten that the duty of the practitioner is to weigh well the question whether it is not possible to obtain by less heroic means, and therefore with less risk to life, results which, although not so striking at first, may in the end, for all practical purposes, be equally good.

APPENDIX TO LECTURE VI

CASE XV. is an undoubted example of incomplete laceration of the bowel produced by prolonged taxis, a subject which has been already fully considered in a previous lecture.

So far as the condition of the gut was concerned, it was a suitable case for the treatment by tapping and suture described in Lecture V.; the patient, however, was so ill (being apparently moribund) that I dared not delay a moment longer than was really necessary before getting her into a warm bed. The treatment of returning the bowel and using a large drain-tube in the ring was therefore adopted with an excellent result.

Since this lecture was originally published in the *Lancet* of October 18, 1890, an objection has been raised to the use of the drainage-tube in the manner I recommend on the ground that its presence may be a source of possible harm. Why harm should result I do not know, nor can I imagine any reason for anticipating it, provided always that the tube is applied and managed in the proper manner, and that suitable cases are chosen for its use.

I have employed the method several times since in exactly the same way, and shall certainly continue to do so in cases of the right kind, for I have seen nothing but good come of it up to the present time.[1] One caution in connection with the matter which I omitted to mention in the lecture as originally published may be usefully referred to here. The tube should, as I have said, be large ; it should also be *without lateral perforations*, as it lies in contact with the peritoneum, which is pretty sure to become locally inflamed ; if, therefore, any lateral perforations exist, the lymph poured out is prone to mould itself in these openings so firmly that the removal of the tube may be impossible without considerable force, which, especially when the tube has been *in situ* for some time, leads to oozing of blood. This difficulty is all the greater when lymph, as generally happens, also travels a short distance up the interior of the tube and unites with any of the sucker-like processes which pass in through the lateral perforations. A tube without these perforations rarely sticks at all ; and if it does so a little at the first attempt at withdrawal, a few gentle jerks instead of continuous traction will release it. It must also be borne in mind that the tube is

[1] February 1893.

I

placed in the belly solely for the purpose of carrying off any discharge, which may be decomposed and septic, from the damaged bowel and peritoneum, and is not intended to drain the soft parts lying immediately around it. The absence of lateral perforations is therefore, to some extent, a safeguard against the contamination of the soft parts in which the tube lies, as no escape from it into these parts can take place whilst the discharges are passing along its canal.

The tendency of peritoneal lymph to fix perforated tubes is, of course, familiar to all surgeons who are much concerned in abdominal operations; and I only mention the fact here for the benefit of those who may not have had experience in the matter.

When gangrenous gut is left unreduced after herniotomy, with a view to the formation, for the time being, of an artificial anus, there appears to be considerable difference of opinion amongst surgeons as to the desirability or not of dividing the stricture at the point of strangulation.

For my own part, I think it is far better that the stricture should be divided—and, indeed, be divided freely. Such free division, in the first place, not only affords comfort to the patient by relieving the continuous pressure on the constricted parts, but also obviates, to a great extent, the occurrence of peritoneal irritation, the symptoms of which are sometimes so nearly akin to peritonitis that the diagnosis between the one condition and the other is very difficult; in the second place, it contributes very markedly to the speedy spontaneous closure of the artificial anus subsequently, for after free division of the stricture, the protruding parts, when the inflammatory irritation and infiltration have to some extent subsided, show a much greater tendency to retract towards the belly than when the stricture is left untouched ; and so more certainly and with less delay is effected the withdrawal of the spur, which is the great bar to speedy spontaneous recovery in these cases.

Although the paragraph at the conclusion of the lecture, referring to the resection of portions of the damaged bowel in these cases was written as many as three years ago, I see no reason for altering the opinion then expressed. I have, since the time mentioned, had to deal with cases in which the condition of the gut was very serious, but in neither of them has the general condition of the patient been sufficiently good to justify the adoption of this radical plan of treatment; and I still feel that the coexistence of a condition of the gut hopeless enough to require resection, with a state of the patient which is good enough to justify the operation, must be very rare.

These cases are always urgent and critical, otherwise the local conditions could hardly be bad enough to suggest the propriety of this proceeding; the main onus, therefore, that falls upon the surgeon is, in the first instance, the *immediate saving of life.*

Under such circumstances, no justification, in my opinion, exists for attempting an ideal result, such as that which follows resection when successful, if by the attempt at its attainment the patient's life is placed in greater jeopardy than would follow other measures which, although less perfect in themselves, may, in the first place, with greater certainty serve the immediate purpose for which they are performed —viz. the saving of life; and secondly, may, after all, ultimately leave the patient in as good a position as if resection had been successfully accomplished.

Again, it must never be forgotten that the gut in these cases cannot be in a condition actually favourable for rapid and sound repair after resection; for this reason only, apart from any other, the success of the operation must be doubtful, even in the hands of those of us who are constantly dealing with the peritoneum and intestines in the course of practice.

So far, therefore, as the ordinary practitioner is concerned, there can at least be no manner of doubt that the formation of an artificial anus should be preferred to resection.

Indeed, for my own part, although I shall, if the suitable case presents itself, perform the operation of resection of the damaged or gangrenous gut with perfect confidence, I shall at the same time (in the words of an astute surgeon, used in reference to a different question [1]) 'not seek the opportunity with ' any hasty view either that I should, on the one ' hand, by a successful result tend to establish the ' propriety of the operation, or, on the other, by ' failure, be deterred from further trial.'

[1] Mr. Bransby Cooper on 'The Extirpation of an Ovarian Cyst.' *Medico-Chirurgical Trans.*, vol. xxvii. p. 76.

LECTURE VII

*ON MEDIAN ABDOMINAL SECTION IN THE TREAT-
MENT OF STRANGULATED AND UNSTRANGU-
LATED EXTERNAL ABDOMINAL HERNIA; WITH
REMARKS ON THE CAUSES OF THE WANT OF
UNIFORMITY IN THE SYMPTOMS CAUSED BY
STRANGULATED OMENTUM*

(Not previously published)

SYNOPSIS—Introductory remarks—Median abdominal section entirely
unnecessary for the radical cure in ordinary cases—Desirability of
performing the radical cure if a hernia is found during abdominal
section for other purposes—Laparotomy never likely to be generally adopted in strangulated hernia, being only necessary in rare
cases—Ordinary herniotomy simple and successful—Advantages
claimed for the major operation—Objections to its performance—
Treatment of strangulated hernia by median abdominal section
rarely safe, excepting when combined with ordinary herniotomy—
Supposed advantages of the major operation not enough to justify
the extra risk—Relative insensitiveness of peritoneum forming
hernial sac compared with that forming general peritoneal cavity—
Abdominal section indicated in cases in which some other intra-
abdominal condition exists which could at the same time be dealt
with—Occasional occurrence of cases in which median abdominal
section is unavoidable—Clinical example (No. XVII.), acute intestinal obstruction in a case of double hernia—Abdominal section—
Remarks on the case—Reason for performing abdominal section in
the first instance rather than herniotomy—Reference to another
case under author's care in which the major operation was considered and rejected for reasons stated in favour of herniotomy—
INCIDENTAL REMARKS ON THE UNCERTAINTY IN THE SYMPTOMS
PRODUCED BY STRANGULATION OF OMENTUM—Further reference to
Case I, in relation to omental strangulation—Difference between
the symptoms produced in that instance and in another case
referred to—Causes of the acute symptoms in the latter case,

and when occurring in omental strangulation generally—Effects of strangulating healthy omentum by ligature—Exceptional example of acute intestinal obstruction caused by an omental band in a case of inguinal hernia.

It is, I need hardly say, common enough to see persons who are the subjects of two or more herniæ (for instance, two inguinal herniæ and a rupture at the umbilicus form a combination which is not very rare). In such patients intestinal obstruction may arise which, from the nature of the symptoms, it is difficult, if not impossible, to connect precisely with either one of the ruptures.

Under these circumstances it becomes necessary to consider the proper course to follow. Should the herniæ be explored one after the other, or should median abdominal section be at once performed?

A case of this kind recently under my care in the wards, affords an opportunity for the discussion of median abdominal section, not merely in this limited application, but also in relation to the treatment of hernia generally.

In the course of the last few years, the question of the propriety of substituting median abdominal section for herniotomy in cases of hernia generally has been from time to time brought forward, and at the meeting of the British Medical Association in 1891 was honoured with a long discussion.

Very little disinterested consideration of the matter must, I am sure, lead to the conclusion that in the radical cure there is nothing whatever to commend the major operation, as there are several methods of treating the sac and canal by cutting directly down upon the hernia, which are so easily applied and so effectual that nothing material can be gained by formal abdominal section. It is at the same time perfectly certain that the adoption of the more serious operation by any person who is not constantly in the habit of dealing with the peritoneum, and who may be without skilled assistance, must involve considerable danger. During the performance of laparotomy for the removal of an intra-abdominal tumour, or for the treatment of one of the many conditions for which abdominal section is now so commonly employed, it would be clearly proper to effect the radical cure of a hernia which happened to exist, by obliterating the sac and ring from the inside by any means which seemed at the time most suitable, if the local conditions were favourable and the state of the patient good enough to justify the extension of the operation with safety.

In cases of strangulated hernia I cannot believe that median abdominal section will ever be regarded as the routine treatment. Indeed, it must, I am sure, be adopted by the ordinary practitioner in rare and altogether exceptional cases only, if the welfare of the patient is to receive proper consideration.

The time-honoured operation of herniotomy is generally so simple that it can be easily performed by any practitioner, with results which are so good when treatment has not been too long delayed that in reality it leaves little to be desired.

The main advantage claimed for the median operation is, that the reduction of the hernia by traction from above is more easily effected than by pressure from below, as happens in the herniotomy commonly practised.

In the advantage thus claimed there would be some merit, if adhesions in the sacs of strangulated herniæ were only rarely found, if omentum only had to be dealt with, and if, when gut exists in the sac, it were always sufficiently healthy and sound to justify the amount of traction necessary for its reduction without previous careful examination.

No demonstration is needed to prove the danger of dragging at a hernia from above composed of gut which may be adherent, injured possibly by unwise taxis, and, worst of all, perhaps, on the point of gangrene, or adherent and ulcerated, it may be, at the seat of constriction. Moreover, every surgeon knows that the fluid so commonly found in the sacs of these herniæ is sometimes dark and ill-smelling when the strangulation has been long existent, although no gross lesion may be apparent in the gut itself.

In herniotomy the fluid contents escape upon

opening the sac (I suppose that no rational surgeon in these days would attempt the reduction of a strangulated hernia without opening the sac), and the bowel itself is thoroughly cleansed before its return into the belly.

In withdrawing the hernia from above, the object to be attained by performing abdominal section, the gut returns into the abdomen uncleansed, and is followed by the fluid contents of the sac, in whatever condition they may be.

Surely this can bring no good to the patient.

It may, of course, be here said that if a piece of tainted bowel or a little stinking fluid should pass up into the belly, flushing with warm water will cleanse the gut and remove the fluid; possibly it may be so, but let it be noted that it is generally far more easy to *soil* the peritoneum than *to cleanse it*.

I confess it seems to me almost impossible that any surgeon who has had much to do with the kind of cases of strangulated hernia with which we, as hospital surgeons, have to deal, could seriously propose to substitute in a general way median abdominal section for herniotomy, when the state of the parts commonly present in the sac is considered.

In fact, it is, I think, clear enough that the reduction of a strangulated hernia cannot be effected with safety by median abdominal section, unless the sac is opened in the ordinary way at the same time, to allow of its contents being examined and cleansed prior to their withdrawal into the abdomen.

Such being the case, the advantage obtainable in reducing the hernia by traction from above is not sufficient to justify a surgeon, in any but rare and exceptional cases, however skilled he may be, in performing abdominal section under these circumstances.

I am aware that it may be said if the gut be healthy and only recently strangulated that abdominal section is, after all, not really much more severe than herniotomy, seeing that the peritoneum is laid freely open in each case; but, apart from many other considerations, it must not be forgotten that the peritoneal sac of a hernia, although continuous with the general peritoneum, is generally much less susceptible to septic influences, and, in fact, is generally altogether less delicate than healthy peritoneum inside the belly, especially in old herniæ when trusses have been long worn.

Should a patient, coming under treatment for strangulated rupture, also have an abdominal tumour (*e.g.* ovarian), apparently easily removable, it may perhaps be right to perform abdominal section with a view to dealing with both conditions, if the constitutional state were good enough to justify the proceeding, which is by no means probable.

Even in cases like this, however, the prudent surgeon will, as a rule, I think, do well to be content to deal with the critical condition first by performing herniotomy, leaving the more extensive operation for a later date.

In saying this, I am not forgetful of the case to be described in Lecture VIII., in which the twisting of the pedicle of an undetected ovarian tumour undoubtedly caused death. In that case, had the tumour been diagnosed, its removal at the time of operating upon the hernia would have been out of the question, as the patient was nearly moribund.

Although, for the reasons just stated, I am strongly of opinion that any idea of substituting median abdominal section *in a general way* for herniotomy should be dismissed, there will occur now and then cases of an exceptional nature in which laparotomy should be preferred to the minor operation. The following case is a good illustration of this point :—

CASE XVII.—*Acute intestinal obstruction; abdominal section; Richter's hernia; herniotomy; death from influenza on sixth day after operation.*

A. R——, aged 32, a butcher, was admitted into the Fitzwilliam Ward under my care on January 18, 1892.

He stated that he had been in perfect health till four months previously, but since that time had suffered frequently from 'colicky' pain and wind in the stomach. Three days before admission the pains set in more violently than usual; he therefore took some purgative pills which acted freely, leaving him apparently quite well. The pain came on again thirty-six hours later, and was followed almost immediately by violent vomiting, which recurred frequently up to the time of admission.

He had been ruptured in both groins since he was ten years old, but the ruptures had never given trouble, as they could always be easily kept up by a truss. The bowels had not acted for forty-eight hours.

The patient was certain his illness had nothing to do with the ruptures, as he was positive they were in their usual condition.

On admission.—The man was obviously very ill. The expression was 'abdominal.' The pulse was small and quick, the temperature subnormal, the tongue moist and not much coated. The breath was foul and stinking. Considerable crampy pain, quite clear of the groins, extending over the upper umbilical region, was complained of. There was frequent retching, and a little stinking semi-feculent vomit was brought up.

The abdomen was soft, and only slightly tender to the touch ; the muscles moved freely during respiration.

In the cæcal region the tenderness was more marked, and an indistinct fulness, or rather unnatural muscular resistance, could be felt upon careful palpation. In both groins the abdominal rings were large but apparently free ; on coughing a hernia could be felt to impinge on the finger placed in either of the rings. This impulse was less manifest on the right side, but nothing like an unreduced or strangulated hernia could be found in the canal.

Soon after admission the patient's condition improved considerably, and all vomiting ceased; it was therefore decided upon consultation to treat the case as if he were merely suffering from colic, &c.

The vomiting, however, soon returned, and abdominal section was performed on the 20th.

Upon introducing the hand into the belly the gut was found to be fixed in the right internal ring, or rather in the parts just above it, which seemed pulled up towards the abdominal cavity. As it did not appear safe to reduce this rupture from within before it had been examined, the sac on the right side was laid open in the ordinary way. The stricture was divided from below, and after the absence of adhesion had been ascertained the gut reduced itself spontaneously.

On examination of the returned gut the mark of the stricture upon the serous surface was found to include a circular patch on one side of the bowel (Richter's hernia) which involved about two-fifths of the circumference only. The patient progressed rapidly. No further vomiting or other unfavourable symptoms occurred until the 23rd, when he fell a victim to the prevailing epidemic of influenza, and died of pleuro-pneumonia in twenty-four hours.

In this patient abdominal section was plainly indicated. The existence of hernia on both sides, the absence of any positive indication of strangulation in either groin, the muscular resistance over the cæcal region, and the general character of the pain, together with the firm conviction of the patient that the symptoms were entirely unconnected with the ruptures, were the main points which guided me

in my decision. The fatal result was purely accidental, and in no way affects the propriety of the
treatment.

This is the only instance in which I have performed abdominal section for strangulated hernia in
the groin. In another case, which occurred some
years ago, I considered the question of opening the
belly in the middle line, but ultimately decided upon
herniotomy, with a result which quite justified the
choice.

The patient was a healthy young man, twenty-
eight years old, whom I was called to see, with intestinal obstruction. There was no obvious indication
to the touch of any external hernia, but he had
suffered from a rupture in the right groin when an
infant, for which he had worn a truss till he reached
the age of fourteen or fifteen, after which it had not
been used. He had for several years suffered, after
any great exertion, from crampy colic-like pains
about the lower part of the belly, and he was
generally constipated in habit. The attack for which
I saw him commenced with 'colicky' pains after
lifting some heavy weights. These pains seemed to
commence in the right groin and then become
general, finally concentrating themselves about the
navel, from which they at times seemed to drag
towards the groin. Although no obvious tumour
like a hernia could be felt, pressure over the right

internal ring caused some flinching, and was said to produce also a dragging sensation from the umbilicus. Bearing in mind the existence of the rupture in childhood, I concluded that the symptoms were very possibly due to a small hernia (partial enterocele or merely omentum) into the upper end of the old sac. In spite, therefore, of the general resemblance of the symptoms to those of strangulation by an internal band, I explored the site of the old rupture, and found a very small elongated sac containing, tightly nipped in its upper end, a tiny piece of recent omentum. The stricture was divided, and the little mass quickly returned into the belly. Had there been evidence of a rupture having previously existed *on both sides*, I think abdominal section would have certainly been indicated, for the pain caused by pressure over the groin I have seen in cases of intestinal obstruction produced by bands and other causes quite unconnected with external hernia.

The case affords a good demonstration of the wisdom of the old surgical rule to which I have before referred, that in all cases of intestinal obstruction occurring in a patient who is or has been the subject of hernia, the first indication, in the absence of positive evidence of intra-abdominal causes, is to explore the site of the rupture.

A question of great interest, and one which has

been much discussed from time to time,[1] is the singular want of uniformity in the symptoms caused by strangulated omentum, especially in relation to intestinal obstruction.

In Lecture I. I have described in detail a case in which a very large mass of omentum in an inguinal hernia was tightly strangulated and on the point of gangrene, in fact at one spot the gangrenous process had actually commenced; yet no symptoms of any kind pointing to obstruction of the bowel occurred, the action of the bowels being quite natural. In the case now mentioned the symptoms of obstruction were, on the contrary, complete, the vomiting was frequent and fæculent, and the patient was rapidly becoming exhausted (so much so that at the time of the operation grave doubts of his power of recovery were entertained), although only a minute nodule of omentum was involved in the constriction.

The extreme severity of the symptoms were here, I believe, due to a double cause: (1) the omentum only recently strangulated was healthy, and therefore very sensitive; (2) it was clear, from the manner in which the small mass sprang back into the belly of its own accord after the division of the stricture, that there was much tension on its stalk; in other words, that there must have been considerable drag-

---

[1] See, for example, a correspondence between Mr. Holmes and Mr. Rushton Parker in the *Lancet*, 1885, vol. i. p. 680.

ging upon the parts inside the abdomen. In the other case, although the omental mass was very large it was hypertrophied, had been for a long time in the sac, and was on these accounts insensitive. The stricture was, it is true, very tight, but it involved no recent omentum, and on its division the stalk of the herniated mass was loose and long.

Judging from the evidence afforded by these and other cases which have come under my notice, there is, I believe, no doubt that the acuteness of the symptoms in strangulated omental hernia, particularly in relation to obstruction of the bowel, depends upon the existence of one or both of the conditions I have indicated (*i.e.* the recent strangulation of healthy omentum and the existence of tension in the pedicle of the strangulated mass), the extreme condition of fæculent vomiting, &c., being probably due to the co-existence of both, the most important factor being the traction upon the abdominal contents by the tense pedicle.

With reference to the symptoms produced by strangulation of healthy omentum, the fact, to which I refer elsewhere,[1] is worthy of attention—viz. that the ligature of pieces of omentum in herniotomy or other abdominal operations is frequently followed by vomiting and pain, without any evidence of peri-

[1] Lecture VIII. p. 142.

tonitis, which pass off in the course of twenty-four or forty-eight hours.

In one case of intestinal obstruction from strangulated omental inguinal hernia upon which I operated, and which proved fatal from exhaustion, the obstruction was evidently caused by the pedicle of the hernia, which was long, and so thin and tense that it had produced enough pressure upon a piece of gut to make a deep groove across its peritoneal surface. This was an exceptional instance, and the only one of the kind I have met with.

LECTURE VIII

ON CERTAIN UNFAVOURABLE SYMPTOMS OCCURRING
AFTER THE REDUCTION OF STRANGULATED
HERNIA WITH OR WITHOUT OPERATION

(Not previously published)

SYNOPSIS—Introductory remarks—COLLAPSE—(1) In neglected cases, especially those in which the vomit has been fæculent—Treatment—Gastric irrigation—Effect of warm water as a restorative—(2) In perforation of the bowel—(3) In recurrent strangulation: diagnosis of these two causes—Treatment of cases of perforation—Abdominal section the only resource, not often justifiable at the stage usually seen—Author's views as to the moral aspect of certain apparently hopeless operations—Gradual perforations of the bowel not necessarily associated with collapse—Further reference to treatment by large drain-tube in canal—Danger of abdominal strain or unnecessary manipulation of abdomen in cases of rupture after reduction—Condition of abdominal tension, &c., of more importance than the state of the wound after herniotomy—Syncope following first action of bowel after reduction not to be confounded with collapse.

VOMITING — Commonest recurrent symptom — Causes — *Immediate* Anæsthetic—Fæculent material in stomach—Unrelieved strangulation—Recurrence of hernia—Ligature of large masses of recent omentum—Perforation of bowel—*Remote* Ulceration of gut—Inflammation or suppuration in sac—Recurrence of strangulation—Remarks on these various causes, and indications as to treatment—Clinical example (No. XVIII.)—Recurrence of vomiting after herniotomy caused by twisted ovarian pedicle—Remarks on this case—Question of abdominal section—Necessity of careful examination of the belly generally before operating for strangulated hernia—Reference to Case IV. in this relation.

THERE are few cases which give rise to more anxiety in the surgeon's mind than those in which unfavourable symptoms follow upon the reduction of an apparently strangulated hernia, whether by manipulation only or after herniotomy.

Such symptoms are sometimes, it is true, of little gravity ; but they are often of the greatest importance, and a patient's life may at any time depend upon a proper estimate of their severity: not always an easy matter, even with those who have had a large practical experience in the treatment of complications of this kind. A little time, therefore, cannot fail to be profitably spent in the consideration of the most important of these recurrent symptoms, with a view to indicating the lines upon which a fairly reliable estimate of their relative gravity may be made, without which their rational treatment is impossible.

The symptoms I propose to discuss at length are :—

1. Collapse.
2. Vomiting.

I. Collapse

This may be immediate—that is to say, following directly upon reduction—or remote.

In neglected and long-standing cases of strangulation, especially in old people, nature seems to make no attempt at a rally, and the patient gradually sinks, dying of sheer exhaustion and not unfrequently partly from intense pulmonary engorgement commonly supposed upon insufficient grounds to be produced by the anæsthetic.

An interesting form of collapse is that which sometimes occurs after herniotomy in ordinary cases of strangulation in patients who have been vomiting stinking material. The patient rallies after the operation and seems to be doing well, when suddenly, perhaps at the end of twelve hours or thereabouts, after copious vomiting of semi-fæculent matters, collapse or rapid exhaustion sets in; the vomiting recurs again and again, and no sign of improvement shows itself *until the vomit has lost its stinking character.* Then, if the patient is not too far exhausted, improvement commences, the sickness ceases, and all goes well.

This kind of collapse is, as I have said, seen only, so far as my experience goes, when the vomit has been fæculent, or nearly so, and it is due entirely to the violent straining and disgusting nausea caused by the efforts of getting rid of the stinking contents of the stomach.

Such being the cause, the treatment is clear enough—viz. washing out the stomach with warm water until the washings lose all their foulness. This treatment not only cleanses the stomach, but affords a direct general stimulant of surpassing efficacy in collapse of this kind. This latter fact is not yet, I fancy, sufficiently realised by many people.

It is quite certain that the passage of water, as hot as it can be borne, into the stomach, whether by drinking or by artificial means, is the best stimulant that can be used in collapse of any kind, being far better in every respect than any form of alcohol.

The genial warmth which rapidly passes over the whole body in many of these cases after the use of hot water is remarkable, and its effect upon vomiting when present is always beneficial.

The most hopeless form of collapse, it is hardly needful to say, is that which follows perforation of the gut, when from long neglect or other causes its walls have become thinned by ulceration or weakened by gangrenous changes. Perforation from these causes may occur immediately after the reduction of the intestine into the belly or at a later period, generally at the time of the first action of the bowels.

The diagnosis of the condition is sufficiently easy in the majority of cases: the rapid change in the patient's state, the almost sudden abdominal tension, the pain with or without vomiting, and at first the subnormal temperature, are symptoms which, taken together, can scarcely be misinterpreted.

In cases in which the hernia becomes again strangulated soon after reduction, collapse frequently follows directly upon the re-strangulation, *but it is always preceded by vomiting.* In perforation of the bowel, on the other hand, the *collapse occurs before the vomiting* ; moreover, the general peritoneal distension, which occurs immediately upon the perforation, is wanting in the other case.

The treatment of these cases is a grave and difficult question. Medicinal measures are of course useless, and it is quite certain that if no surgical relief can be given the patient will die.

The only hope, therefore, if there be any hope at all, lies in abdominal section, the damaged gut being dealt with as the condition of the patient and the local circumstances best permit.

Abdominal exploration under these conditions generally offers a hope which is so forlorn that the proceeding is rarely justifiable, as the patient's state is usually so critical.

I have once only explored the abdomen in such a case, and I have never ceased to regret it, although the operation was done at the urgent request of the patient and his friends. I felt that the case was hopeless, but, under the united pressure of those concerned, it was difficult to absolutely decline to do anything.

If abdominal section could be done at the time of perforation, or *immediately afterwards*, there may perhaps be a prospect of saving life ; but in all the cases I have seen the time for active interference has passed, the patient being clearly moribund, or so nearly approaching that state, that any operation has been entirely out of the question. Nevertheless, when the symptoms are distinct, and the patient's general state sufficiently good to hold out any prospect of recovery, however small, it is undoubtedly the surgeon's duty to explore the abdominal cavity and do the best he can, provided always that the patient or his responsible friends have been made fully acquainted with the gravity of the situation.

It is in connection with cases of this kind that opinions are occasionally given in favour of operating, because the patient, if left untouched, must inevitably die ; and because an operation, if performed, at least cannot make matters worse, although the chances of its effecting any good are practically nil.

Against recommendations in favour of surgical interference in any case upon such grounds as these I cannot too strongly protest.

Surely the justification for an operation lies in the prospect of its achieving some good for the patient —not in the supposed impossibility of its doing harm. Indeed, for my own part, I do not know how it is possible to say that an operation in circumstances like these we are considering can do no harm, although the ultimate death of the patient may appear certain ; for it undoubtedly may, and probably will, in these desperate cases shorten life ; it inflicts unnecessary distress and suffering upon the unhappy patient, and causes needless additional anxiety to friends, whilst it has no rational end of any kind.

In my opinion, any operation performed without some distinct chance of conferring benefit upon the patient is an act of immorality, no matter how certain it may appear to be that the proceeding can do the patient no material harm.

It is here desirable to note that perforation of the bowel is not necessarily associated with collapse, excepting when the general peritoneal cavity is involved. Should the gut be adherent to the parts about it, ulceration through its walls produces only a fæcal abscess, and in cases treated by the large drainage-tube, as described in Lecture VI., no unfavourable symptoms of any kind need arise, as the tube provides for the immediate and direct escape externally of the contents of the gut.

Whilst speaking of this form of ulceration of the gut, it is well to bear in mind that any violent contraction of the abdominal muscles exerts great pressure on the intestines, and in bowel weakened locally by ulceration or other causes tends to bring about perforation or sudden rupture into the peritoneal cavity. I mention this obvious fact as a warning against allowing patients to raise themselves in bed, or perform any action unaided which is likely to cause abdominal contraction, until after the first action of the bowels following the reduction of a strangulated hernia. After the bowels have acted, this point is of less importance, as the intestines are not so much distended, and are therefore less liable to injury; at the same time, if the first motion contains altered blood in any noteworthy quantity, great care should be used subsequently for a week or ten days at least. It not uncommonly happens that the practitioners of no large experience concentrate their attention too much upon the state of the operation wound, so that the condition of the gut itself, a matter of much graver import, is apt to be neglected. It must not be forgotten that the wound may heal by first intention, although the bowel itself is in a condition of serious disease. By all means let the wound be well cared for, but at the same time let the condition of the belly as to tension, rigidity, and tenderness be carefully noted, for the satisfactory progress of a patient is gauged not so much by the condition of the wound as by the general state of the abdomen.

The feeling of faintness, or actual syncope, which is not very unusual during or immediately after the first action of the bowels subsequent to the relief of the strangulation, should not be mistaken for collapse ; the familiar signs of faintness are present, the patient rallies directly, and there is an entire absence of serious abdominal complication.

II. Vomiting

This is the commonest of all the symptoms which supervene upon operations for the reduction of strangulated hernia; it is also, on the whole, the most important, since it may be the indication that the strangulation still persists, having been unrelieved, or that the hernia has again become strangulated.

Vomiting after the reduction of a hernia is chiefly due to one or more of the following causes :—

1. *Immediate:*

a. The anæsthetic.

b. Fæculent or foul material remaining in the stomach.

c. Unrelieved strangulation.

d. Recurrence of the hernia.

e. The ligature of large masses of recent omentum.

f. The fixation by suture or ligature of pieces of recent omentum to the abdominal parietes if any traction is thus made upon the attached structure.

g. Perforation of the bowel in neglected cases.

2. *Remote:*

a. Ulceration of the gut internally.

b. Peritonitis.

c. Inflammation or suppuration of sac, &c.

d. Recurrence of the hernia.

The vomiting caused by the anæsthetic, although distressing, is perhaps too characteristic to need much comment. By degrees it usually stops spontaneously. The traditional treatment by ice is altogether faulty. It depresses the patient, who is already sufficiently exhausted, it causes a most distressing thirst, and last, but not least, frequently fails to affect the symptom in any way. Hot water sipped frequently is stimulating and much more effectual.

The vomiting caused by stinking material in the stomach usually cures itself by the emptying of the viscus which follows spontaneously ; the nausea, however, caused by the offensive nature of the vomit is sometimes so persistent as to become serious ; and I have seen aged patients die from the exhaustion thus caused.

In such cases the treatment indicated is gastric irrigation, originally suggested, to the best of my belief, by Mr. Frederick Treves. The stomach should be thoroughly and copiously washed out with hot water, or, better still, with a wash composed of boro-glyceride in hot water ($\mathfrak{z}$j. to Oj.). If the vomiting is merely due to residual fæcal accumulation, the washing cleans out the stomach thoroughly, removes the disgusting taste, and benefits the sufferer in a manner which is often remarkable, raising the condition from apparently hopeless prostration to comparative comfort.

When there has been fæcal vomiting previously, washing out the stomach at the time of operating is especially beneficial. This is easily done whilst the parts are being cleansed and prepared for the operation, immediately after the patient is under the influence of the anæsthetic, or, better still, at the conclusion of the operation before consciousness has been fully recovered.

A less familiar cause of vomiting than either of the foregoing is the inclusion in a ligature of large pieces of recently herniated or healthy omentum : a point which is referred to in Lecture VII.

Vomiting caused in this way is not rare, and accounts for the fact—which would not at first be expected—that cases of herniotomy in which omentum has been dealt with very commonly progress less quickly than those in which the sac has contained gut only, in consequence of the pain, irritation, and vomiting caused by the ligature. *Cæteris paribus*, the larger the portion of omentum ligatured the more severe the symptoms generally are. In fact, there are no symptoms which simulate unrelieved or recurrent strangulation so closely as those which sometimes follow upon the ligature of healthy omentum. Fortunately, however, there is no difficulty in producing an action of the bowels by enemata or otherwise, so that no legitimate reason for error in this particular exists.

Inflammation or suppuration of the sac as a cause of vomiting after herniotomy is easily recognised by the state of the parts about the operation wound. In all cases when vomiting is associated with great pain, swelling, tenderness, and tension about the scrotum and neighbourhood, the wound must be opened up and the parts examined and thoroughly cleansed. The use of drainage for twenty-four hours after the operation prevents the accumulation of retained discharges, and so to some extent obviates the tendency to this uncomfortable condition of things.

The inclination with surgeons now is to dispense with the use of the drainage-tube altogether in these operations, but if any part of the sac is left unremoved it certainly should as a rule be used, excepting in infancy and childhood, but it should not be allowed to remain in the wound more than twenty-four hours.

Before laying open the parts care should be taken to ascertain that the symptoms are not due merely to acute orchitis, as sometimes happens, since opening up the parts would then be clearly useless, if not harmful.

It is worth mentioning, perhaps, that the most severe vomiting of the kind under discussion which I ever saw was caused by acute effusion into the sac of a large hydrocele which was stated to have been ' cured ' by injection some years previously.

When the mucous lining of the gut is ulcerated

around the line of stricture, as may happen if the constriction is very sharp and tight, or if the strangulation has been long existent, vomiting occasionally follows immediately upon the first action of the bowels after the reduction of a hernia. The occurrence of altered blood in the motion and a slight rise in temperature, generally followed by a little diarrhœa, are symptoms which, as a rule, render the diagnosis easy.

Vomiting from this cause is much more likely to happen when the motion has been made dry and hard by the administration of opium, a common practice, apparently, with practitioners in cases of strangulated hernia, and one which should be avoided, for many reasons, if possible.

The vomiting usually subsides of its own accord after the bowels have been freely opened spontaneously, or by an enema of olive oil or some other gentle method.

The irritability of the gut, as shown by the diarrhœa, generally subsides without treatment, or immediately yields to a single moderate dose of laudanum.

The above are fairly representative examples of vomiting, without abdominal symptoms indicating serious local lesion. When associated with obvious changes in the condition of the abdomen itself, the symptom, of course, assumes a much graver aspect.

In most of these cases the abdominal symptoms (*e.g.* peritonitis, perforation of the bowel, &c.) are so distinct that the diagnosis presents no difficulty.

The only cases, therefore, which I propose to discuss now are those in which the strangulation has not been relieved by the reduction of the rupture by operation or otherwise, and those in which the hernia has again become strangulated after reduction.

It will be evident, from what I have already said, that vomiting associated with scrotal swelling, pain, and tenderness, is, although suggestive, not conclusive evidence of the recurrence of the hernia, since it may be due to inflammation or suppuration in the sac, acute hydrocele, orchitis, or even cellulitis.

In such cases, before proceeding further, a copious enema of warm olive oil should be given; if free evacuation follows, the existence of recurrent or unrelieved strangulation is pretty conclusively negatived. If no action is thus produced, it is better for the good of the patient to assume that obstruction of the bowel is present, and proceed accordingly.

Vomit tending to be stercoraceous in character, although almost certainly due to obstruction of the gut, may arise from peritonitis only; there is, however, but little practical importance in the diagnosis of this cause in the class of case now being considered, as, under any circumstances, abdominal exploration is clearly indicated.

If the swollen scrotal parts, after being laid open, are found to contain no gut, the finger must be passed into the belly through the ring as far as it will go, in order to explore the canal with a view to the detection of a reduction *en bloc*, or other unnatural condition within reach.

Should any abnormal state of affairs be perceptible to the finger the canal should be slit up, the parts exposed, and dealt with as the discretion of the operator directs. If the ring is free, the question then arises as to which is the better plan of making the abdominal exploration which is clearly called for: should the abdomen be freely opened by a prolongation of the operation wound, or should this be closed up and a formal median abdominal section be made?

The decision upon this point will depend partly upon the condition of the patient, and partly upon the state of the tissues about the operation wound.

If the patient is collapsed, exhausted, or semi-moribund, the method which occupies the least time must be chosen (*i.e.* slitting up the canal). If the patient is in a fair condition generally, the decision will turn entirely upon the local condition of the parts. If the tissues are clean and not inflamed or

septic, the best plan is to open up the belly by extension of the operation wound ; if, on the other hand,
these parts are foul or suppurating, it is undoubtedly
wiser to close them and open the belly in the middle
line, as the danger of septic infection must be less
when the bowels are handled through a new clean
wound in the parietes, than when dealt with through
one which is obviously septic.

The complications which sometimes give rise to
symptoms of unrelieved or recurrent strangulation
are very remarkable and most difficult to recognise,
as the following case will show.

CASE XVIII.—*Strangulated femoral hernia ; herniotomy ; recurrence of stercoraceous vomiting on third day
after operation ; collapse and death ; ovarian tumour with
twisted pedicle found at* post-mortem *examination.*

E. L——, a woman, aged 45, was admitted under
my care on January 22, 1888. She had been ruptured
in the left groin for three years, and had during this
period constantly worn a truss.

On the 18th the bowels had acted freely for the last
time, and on the 19th a ' fresh rupture came on the top
of the old one.' Vomiting followed immediately, and
continued at short intervals up to the time when she
came to the hospital.

On admission.—The patient was an immensely fat
woman, who constantly vomited fæculent matter. Her
face was anxious, the tongue dry and brown, the pulse
feeble and quick, the temperature 99°.

In the left groin was a strangulated femoral hernia,

tender, tense, and painful. The abdomen loaded with fat was prominent, but not notably tense.

Herniotomy was at once performed. The sac contained a considerable mass of old omentum and a knuckle of intensely congested, almost black, gut. The stricture, very tight and sharp at its edge, was freely divided, the omentum removed, and the gut gently returned into the belly.

The sac was ligatured and removed, and the parts brought together as usual.

The patient rapidly gained strength after the operation ; on the following day free passage of flatus per anum occurred, and gave her great comfort. In the afternoon she vomited once, the ejected matter being of a bilious kind only.

On the 24th the bowels acted, and more flatus was passed. She remained comfortable and free from pain till the 25th at 11 A.M., when she vomited stercoraceous material and rapidly became collapsed.

The wound appeared perfectly healthy, but in order to avoid the possibility of overlooking any recurrence of the hernia, I opened it up and found nothing.

At a consultation held an hour later, it was decided that the patient was too ill to justify abdominal section, and she died a few hours afterwards.

Post-mortem examination.—In the lower part of the abdomen was a unilocular left ovarian cyst, the pedicle of which was long and twisted upon itself to the extent of one turn and a half. The twist was recent, and there was no lymph upon the pedicle or about it. Close to the left internal abdominal ring the omentum was adherent by old adhesions. The end of this piece of omentum passed towards the neck of the sac where it had been ligatured and cut off. Five feet above the ileo-cæcal valve was a ring-like mark upon the intestinal peritoneum, where it had evidently been caught in the stricture. At this spot there was an annular ulcer of the mucous membrane.

There was no peritonitis.

It is plain that the ultimate cause of this woman's death was the disturbance produced by the twisted ovarian pedicle, for the ulceration of the inner surface of the gut could hardly have had anything of importance to do with the result.

The existence of the tumour was not suspected till it was seen at the *post-mortem* examination, for, although the usual examination of the abdomen was made before the herniotomy, no tumour was detected ; this was not very remarkable, seeing the extreme obesity of the subject, and bearing in mind the fact that she was too ill to justify any more delay for the purpose of examination than was absolutely necessary. Indeed, as it was, it seemed very doubtful whether she could be got off the operating table alive.

The issue of the case was not, I think, in any way affected by my ignorance of the existence of this tumour ; for clearly, if I had detected it, its removal would have been out of the question at the time of the herniotomy, and impracticable at the time of the sudden recurrence of the symptoms, as the patient was too ill to be subjected to an abdominal section.. Had I, however, been aware of the state of affairs in the belly, I should, as anything more radical was negatived by the condition of the patient, certainly have tapped the tumour when the unfavourable symptoms came on, with a view to diminishing the intra-abdominal pressure in order to give, to some extent perhaps, relief to the intestines, which were of necessity unnaturally compressed, as the tumour was of considerable size. Whether this proceeding would have effected any good I cannot say, but I conceive it to be possible that it might have allowed the patient to rally sufficiently to have made abdominal section justifiable.

One thing at least is emphasised by this complicated case, viz. the necessity of a more careful general examination of the abdomen than is, so far as I know, usually made before operating for hernia; for although, as I have said, no difference in the termination of this particular instance would, I believe, have resulted from a knowledge of the existence of the cyst, it might under other conditions, for instance, if the patient had been in a less critical state, be a matter of vital importance. This last point is well illustrated by Case IV., in which, had I been unaware of the dull area in the abdomen, I might have been led to the conclusion that the symptoms.

resembling those of intestinal obstruction had occurred in connection with a sac containing fluid only, and so might have sent the man back to bed without detecting the internal hernia, a mistake which must certainly have been a fatal one.

LECTURE IX

ON THE RADICAL CURE

(Not previously published)

SYNOPSIS—POINTS FOR CONSIDERATION IN RELATION TO THE RADICAL TREATMENT—PRESENT CONDITION OF PATIENT—Ruptured persons unsound and always subject to risk—Risk not entirely obviated by use of truss—Dangers of inefficient trusses—GENERAL HEALTH—In the radical treatment risk of the operation to be weighed against advantages offered by it—PROBABLE BENEFIT OBTAINABLE dependent on three conditions—viz. age of patient, condition of the hernia, and form of operation—Discussion of these points—TENDENCY TO SPONTANEOUS RECOVERY IN THE YOUNG—Truss often necessary after the radical cure in middle life and advanced age—Operation should not be rejected on this account—Probable results at various periods of life in respect to the necessity for use of trusses—Incompleteness of the results of spontaneous recovery as compared with radical cure—Condition of contents of sac never necessarily negative radical cure—Radical treatment in unstrangulated cases—RISKS OF THE OPERATION—Danger to life infinitesimal in the ordinary run of ruptured people—Risk to parts concerned in the operation, *e.g.* vessels, vas deferens, and testis—Question of removal of testis in certain cases—Radical cure in strangulated hernia—RELATION OF SOCIAL CONDITION OF PATIENT TO RADICAL TREATMENT—Summary of cases suitable for operation—Further reference to fallacy of estimating value of the radical cure solely upon the possibility of dispensing with use of trusses after operation.

IN order to arrive at a proper decision as to the desirability of recommending the performance of an operation for the 'radical cure' of hernia, the following points require careful consideration :—

1. The present condition of the patient with re-

gard both to the hernia and to the state of the general health.

2. The probable benefit obtainable from the treatment.

3. The risks immediately connected with the operation, especially in comparison with the danger of strangulation, obstruction, or injury, which may exist in any given case.

4. The social condition and occupation of the patient.

_____

THE PRESENT CONDITION OF THE PATIENT

It must at once be conceded that a person suffering from rupture is essentially unsound, since he is the subject of a defect which in the majority of cases interferes to some extent with the ordinary habits of life, unless they are of a very sedentary kind. Further, a ruptured person is always in a certain amount of danger from strangulation, blocking or injury of the hernia. The *amount* of danger from these sources necessarily varies in different cases, but it is to some extent present in all. Unfortunately, the fact that the hernia may be continually controlled by a properly-fitting truss does not entirely neutralise this danger, although it may render any disaster improbable. One of the commonest causes, as every hospital surgeon knows, of strangulated hernia in persons engaged in manual labour or in any pursuit involving much strain or exertion, is the breaking or sudden displacement of a truss, an accident which will occasionally happen, however well fitting the instrument may be.

The danger connected with a rupture necessarily increases in direct proportion to the inefficacy of the truss, for a hernia over which a truss is worn, but which at times comes down behind the instrument, is a source of continual serious risk.

An inefficient truss is not only useless, but it is in itself dangerous. In cases of irreducible hernia, continual risk naturally is present varying in degree with the size and rigidity of the ring, the amount of gut in the sac, the extent of adhesion between the gut and the sac or omentum; the risk being probably greatest when portions of the bowel itself are matted together. This risk is still further increased by the wearing of a strong truss over the neck of the hernia, which seems to be a fairly common practice, with the object of preventing a further descent of gut. After due consideration of this part of the question from all aspects, it must clearly be admitted that every person suffering from hernia runs continually a certain amount of risk which varies greatly in degree in different cases. Whilst it is infinitesimal perhaps in a patient of sedentary habits, with a small reducible rupture continually controlled by a well-fitting truss properly worn,[1] the danger becomes gradually more grave in proportion to the activity of the subject,. the inefficiency of the truss, the size of the hernia, its irreducibility and the existence of adhesions to the sac, and between different parts of the bowel.

With reference to the patient's general health,. there is not a great deal to be said. When proposed purely as a matter of expediency, the operation for the radical cure is rendered quite unjustifiable by any

---

[1] The views as to the management of trusses sometimes held by patients, even amongst the well-educated, are certainly curious. It is, for instance, not uncommon to find the truss in an ordinary case of inguinal hernia habitually worn over the iliac region quite above the rupture. Some time ago I was consulted by an elderly gentleman on account of a small, easily-reducible, scrotal hernia. A truss was ordered, and properly fitted. A month later he again called to express his gratification at the result, as he had derived so much comfort from the appliance, especially since he had ' got into the habit of putting it on in the right place.' On examination, I found he had placed it about midway between Poupart's ligament and the umbilicus. I explained, with much care, that for various reasons the truss would be of more use if it were worn over the opening through which the hernia came down, and adjusted it myself after reducing the rupture. About a week after this, the patient called for a moment to say that, although my reasons for placing the truss as I had done were, no doubt, from my point of view, very proper, he had felt so much more comfort from wearing it in the manner he had adopted himself that, at all events for the time being, he had determined to use it in his own way. The following is a more remarkable instance :—A person of acknowledged position in the intellectual world consulted a surgeon in London concerning a small reducible rupture. A truss was ordered, and fitted in the usual way. Some weeks afterwards, the patient again presented himself and expressed his approval of the instrument. As the truss was not being worn at the time, the desirability of its constant use was strongly urged for reasons then fully explained. The patient appeared to understand the position of affairs, but on leaving, said, ' I think I do not care to wear the truss always, but I shall have it at hand, and you may rely on my using it *when necessary.*'

such condition as albuminuria, general weakness, or any other state which may in itself be a source of risk, or which would be held to negative any other operation of expediency. When proposed as a preventive measure in cases in which the chances of strangulation, injury, &c., are strong, the state of the general health, although not perfect, may be ignored so long as there is no actual organic disease likely to give rise to immediate danger from the operation or anæsthetic. In other words, when the general health is not perfect, the question of the propriety of the radical treatment must be decided by weighing the risk continually present from the hernia against the probable danger of the operation. If the continual risk appears to be in excess of the danger immediately connected with the operation, the radical cure should certainly be advised; if, on the other hand, the reverse is the case, then the operation should be entirely rejected.

The Probable Benefit obtainable from the Treatment

In a healthy person, three main factors are concerned in the prospect of the ultimate success of the operation for the radical cure :—

(*a*) The age of the patient.

(*b*) The condition of the hernia.

(*c*) The nature of the operation.

The Age of the Patient.—The younger the patient, *cœteris paribus*, the more certain is the prospect of a perfect result. This is partly due to the notorious fact that young subjects recover with ease and rapidity from all operations, but it principally depends upon the tendency to the spontaneous closure of the unnaturally patent hernial canal in the ordinary process of growth, provided that the hernia is prevented from descending through it. It is well known that in children, if a truss can be adapted which will prevent the descent of a hernia continuously for several years, the objective evidence of the existence of the hernia may, in some cases, disappear, and what is commonly called a 'cure' result; not, I believe, in consequence of the retraction of the sac as some people have imagined, but because the pillars of the ring come together and present so much obstacle to the descent of the rupture that, under circumstances of ordinary pressure, it may be kept up altogether.

This is of great importance in connection with the radical treatment in the young, as the closure of the ring by operation is, in them, such a purely subsidiary matter that it is a question whether in many of such cases it is necessary at all, a point which will be again dealt with later on.

The tendency to spontaneous closure of the ring is present in a gradually diminishing degree in all subjects up to the age of 20 or 25, after which for practical purposes it may be said to cease.

From the age of 25 through middle life the chances of complete success, although less than in the earlier periods, are strong, provided a proper operation is performed.

From the age of 55 upwards the operation must, as a rule, be regarded only as a means of enabling a patient to wear a truss with safety and comfort, for, although in a few cases the ' cure ' may be complete, it must not be anticipated with any confidence. Moreover, the perfect result (*i.e.* if an attempt is to be made to relieve the patient entirely of the necessity for using a truss) can only be obtained by a more complicated and, in some respects, more severe operation than when the object in view is only to place the patient in the position of being able to prevent with certainty the rupture from re-forming by the use of a suitable truss, in cases in which, before the radical treatment, the truss acted with uncertainty or was altogether inefficient.

Like many other surgical proceedings in themselves excellent, the reputation of this radical treatment of hernia has suffered greatly from the over-confidence of its advocates, who have often been led to promise results too good to be uniformly possible of attainment.

One of the commonest of the objections which have thus arisen to the practice of this so-called radical ' cure ' is that a *perfect* result follows in *a certain proportion only* of the cases treated, inasmuch as many patients after the operation are not relieved of the necessity of wearing a truss.

As a matter of pure criticism this objection is obviously to some extent sound, but it can hardly be held to afford any real obstacle to the general adoption of the treatment for the reasons now to be discussed, with which the great majority of surgeons must, I think, agree.

So far as I am able to judge from my own experience, which is not a small one, and from what I see and read of the practice of others,[1] a properly devised operation will in young subjects who are still growing, *i.e.* when the age does not exceed 20 or 25, be followed by results which may be regarded as perfect, since no truss will be necessary. The

---

[1] The radical cure of hernia demonstrates, better perhaps than almost any other condition, the futility of attempting to estimate the general value of a treatment from the practice of any individual surgeon.

The methods used are so numerous and, in many respects, so diverse, the views as to the legitimate scope of the treatment vary so much with different operators, and, most important of all, the discrepancy of opinion held by different surgeons as to the circumstances which justify the operation is so great, that it is quite impossible to form any reliable judgment of the real merit of the treatment, unless it is considered in the most comprehensive manner.

patient will, in fact, be sound in the ordinary meaning of the word, fit to follow any occupation, and acceptable for service in the army or navy.

At the termination of the period of growth the natural tendency to closure of the rings is very small excepting in extremely recent cases ; the hernia will probably be larger, the rings more stretched, and from circumstances which are practically unavoidable greater strain will from time to time be thrown upon the parts. Moreover, complications of various kinds may arise rendering the complete cure by operation more difficult and uncertain. After this period, therefore, a truss in some cases will almost certainly be necessary if the hernia has been in existence long, if it is very large, and especially if the abdominal parietes are generally weak and flabby, as is sometimes the case.

From the age of 25 to 30 I should say that very few cases ought to occur in which a truss is needed after the radical treatment. From 30 to 40 years of age, probably not less than 15 or 20 per cent. should wear a truss, not so much because it is *absolutely necessary*, but as a precautionary measure. From the age of 40 to 60, it will be found that 50 per cent. or more of the patients submitted to operation will certainly require a truss, as a matter of precaution if not of actual necessity. In persons over 60 years of age, excepting perhaps in a few cases, the radical treatment must not be expected to render the use of a truss unnecessary.

In considering the large proportion of patients in which I have stated that the use of a truss after the radical treatment is either advisable or necessary, it should be remembered that the instrument in these cases is worn with comfort and *certainty* as to its efficiency, and not with discomfort, difficulty, and *uncertainty*, as is so often the case in persons suffering from hernia.

Moreover, after the radical cure the truss is worn merely to afford support in order to prevent the *new development* of the rupture under conditions of great pressure, and not as when worn in the ordinary way *to keep up a hernia which has already formed*. The difference in these conditions must be at once obvious to all.

It therefore seems to me that an operation which, if it does not always make the patient independent of a truss, at all events renders the use of the instrument comfortable, certain, and safe, whereas before it may have been intolerably irksome, inefficient, or even distinctly dangerous, requires no argument to justify its recommendation, seeing that the benefit which it affords, although not uniform in degree in all cases, is on the whole undeniable.

If my statements with regard to young growing

M

subjects are true, as I have no doubt they are, it follows that the chances of complete success in them, if it is not absolutely certain, should be very nearly so, whilst the operation in competent hands entails no material risk. In healthy patients of this class, therefore, the radical treatment may (indeed, I believe, *should*) be recommended without hesitation.

I am aware in strongly expressing this view that I fairly lay myself open to the following question: Why urge the performance of the radical cure so decidedly in very young subjects, when it is admitted that the persistent use of a truss will, in some cases, if the hernia is kept up continually for a long period, allow the ring to close, the neck of the sac to contract, and the rupture, so far as any objective evidence is concerned, to disappear entirely?

The answer to this question is as follows: This natural 'cure' (so called) allows of closure of the ring but does not obliterate the sac, as the operation certainly does if it is properly carried out. The effect of the non-obliteration of the sac is to make the liability to hernia in adults who have suffered from hernia in infancy very considerable. For it will be found upon inquiry that a large proportion of adults who develop hernia have been the subjects of rupture in infancy or childhood. The obliteration of the sac by operation in these cases, in my opinion, removes this special liability to the affection in later life.

Further, the proper adjustment and wearing of trusses in the very young are often very difficult matters, as is shown by the fact that in a considerable number of such patients the instrument will not be found over the seat of the rupture at all, being in many cases of the inguinal kind—for instance, worn nearer the umbilicus than the groin!

In fact, a goodly number of the so-called cures which occur in these little patients, for which the truss receives the credit, are in reality quite independent of the instrument, being entirely due to the strong natural tendency to the closure of the ring and canal.

THE CONDITION OF THE HERNIA.—Setting aside any peculiarity of situation, an ideal case of hernia for the performance of the radical cure is one in which the rupture is *complete, small,* and *recent*; in other words, one in which the gut or omentum has actually left the proper abdominal cavity, in which the peritoneum is healthy, in which the contents of the sac are not adherent or otherwise unnatural, and in which the margins of the opening through which the rupture has passed have not become attenuated from the effects of long-continued stretching, so that their natural resiliency has not been destroyed. In such a case a perfect result from the radical treatment should be certain at any period of life.

Bubonocele (incomplete inguinal hernia), excepting when it has become nearly complete, is of all forms of inguinal rupture the least satisfactory for the radical treatment, as no proper obliteration of the sac can be effected without splitting up the structures forming the wall of the inguinal canal, which I think should never be done unless absolutely unavoidable, for reasons which will be stated in considering the operation itself.

Indeed, for my own part, I doubt whether an incomplete hernia of this kind, when existing merely as a slight bulging of gut at the internal ring, should be subjected to the radical operation at all, unless some very special reasons exist for the treatment.

Incomplete femoral hernia is not adapted for the performance of the radical cure, as the operation can only be properly performed by abdominal section, which is hardly justifiable under ordinary circumstances.

These views concerning incomplete hernia, especially of the inguinal variety, are, I know, not in accordance with the opinion of some surgeons; nevertheless, I am confident that they will in the long run be found to be sound.

Large and long-standing ruptures are naturally

less likely to be followed by a perfect result than those which are recent and small, mainly because the abdominal parietes about the ring have become so stretched and thin that after the obliteration of the sac they afford no support of any useful kind, since their resilience is nil, and the approximation of the margins of the opening is in the majority of cases only of temporary benefit. Sometimes, however, even under these unfavourable conditions, a complete ' cure ' may be effected if the right operation is performed.

In old, very large ruptures, which have been irreducible or unreduced for long periods—*i.e.* those in which the contents of the sac have been for a long time continuously outside the proper abdominal cavity (*e.g.* large irreducible scrotal or umbilical herniæ), a considerable obstacle to the production of a good result arises from the fact that when large portions of the abdominal contents have remained outside the belly for long periods continuously, the abdominal cavity appears to contract, its general capacity being thus diminished, so that very great force is necessary in order to return the extruded portions of intestine. The result of this is that the pressure upon the obliterated remains of the sac and upon the ring may be so great, during the first twenty-four or forty-eight hours after the operation, that the hernia may be in danger of bursting through the recently occluded parts. That this accident should actually occur is extremely unlikely, but, as it happened in a case of my own some years since, it is worthy of mention.

I know of no condition affecting the inside of the sac or its contents which renders the performance of the radical cure either impossible or improper, or which need in any way impair the result. Complications such as adhesions between the sac and its contents, or between different portions of the contents themselves, may make the operation more difficult, and the risk, so far as the treatment itself is concerned, perhaps a little greater; but I have never met with any difficulty of this kind which could not with a little care be successfully overcome. The large redundant masses of hypertrophied omentum frequently met with in old irreducible ruptures, adherent or non-adherent, are often of some advantage, as they can be utilised, as I shall endeavour to show, in bringing about the closure of the ring in cases in which it would be otherwise impracticable.

Risks of the Operation

These are of two kinds. (*a*) Those immediately affecting life, and (*b*) those affecting the integrity of the testicle or other parts directly concerned in the operation.

(*a*) The Risk to Life.—This, if ordinary care be

taken in the selection of healthy patients, and in the practice of surgical cleanliness, is, I believe, infinitesimal. In a large experience I have seen no fatal case, and in one instance only have I felt any real anxiety on this score. In this case, which occurred to me recently, suppuration and extreme hyperpyrexia caused by the sloughing of the ligatured end of the sac produced an apparently critical state of things for a few hours. All, however, subsequently went well. The case was one of double inguinal hernia, and the largest I have ever operated upon, the ruptures had for a long period been unreduced, and the tension upon the parts after each operation was enormous; the treatment was urgently called for, as strangulation was constantly threatened. The case, therefore, was of an exceptional character.

That some slight risk must be run in these operations is obvious, as the patient must be deeply anæsthetised for periods varying from twenty minutes to an hour; so far, however, as the treatment itself is concerned, it seems to me that in the hands of any surgeon accustomed to the operation, and who exercises ordinary discretion and skill, the danger amounts to nothing. Certainly it is not more than the majority of persons suffering from pronounced rupture are running daily, and not more than any traveller exposes himself to during a journey in an express train.

(*b*) Risks to Parts immediately concerned in
the Operation.—Although important vessels lie in
the immediate locality of the parts dealt with in the
radical cure of some herniæ (*e.g.* femoral), I do not
suppose they can be in the least danger of injury,
excepting at the hands of grossly careless or ignorant
persons, or in cases of abnormality so erratic and
rare as to be almost beyond the range of possibility.
The only structures liable, so far as I know, to injury
occasionally in practising the radical cure are the
spermatic cord and testicle in some cases of inguinal
rupture, especially of the congenital variety. In such
cases division of the vas deferens has accidentally
occurred, and the injury to the testicle resulting from
indiscreet attempts at the removal of the whole sac
may cause serious impairment or even destruction of
the organ. Such occurrences must, however, be of
the rarest in competent hands; and it is well to bear
in mind with regard to this part of the matter, that
neither the actual loss of one testicle nor its functional annihilation, as would, of course, happen after
division of the vas deferens, has necessarily any ill
effects upon the vitality or procreative power of the
patient, provided, of course, that the other testis is
healthy. In fact, so lightly is the loss of one testicle
regarded by some, that its removal is advocated in
order to facilitate the operation for the radical cure
in certain cases. This I cannot regard as anything

but bad surgery, unless the testicle is useless, diseased, imperfectly developed, or undescended, since the removal of the healthy organ is in my opinion never *necessary* for the proper carrying out of the treatment, and I have not yet met with a case in which its removal could affect the final result in any way whatever.

Up to this point my remarks must be taken to apply exclusively to unstrangulated hernia.

In cases of strangulated rupture, the propriety of performing the radical cure depends mainly upon two conditions—viz. the state of the contents of the sac, and the general condition of the patient. If the sac contents are sufficiently sound to allow of their being returned without risk into the abdomen, the radical operation should be undoubtedly carried out at all periods of life, unless the state of the patient is too critical to safely allow of the prolongation of time necessary for its performance.

The Social Condition and Occupation of the Patient.

These are necessarily important factors, of which due account must be taken in deciding as to the propriety of undertaking the radical treatment.

It is, for instance, clear that a hernia which perhaps involves no risk in a wealthy person, who is in a position to take all necessary care, and to whom the

expense entailed by the use of the best instruments
is of no consequence, may in a labourer be a source
of constant danger.

Again, a sedentary life such as that led by many
business men and clerks, may cause little or no
danger from a rupture which would be a frequent
source of risk in an active subject, like a soldier,
hunting man, or athlete.

Further, it may often be justifiable to subject a
young man to the very slight risk which the radical
cure involves, in order to make him practically
sound, who is unmarried and has himself only to
consider, whilst it may be open to serious question
whether it would, under the same circumstances, so
far as the rupture is concerned, be right to subject
a married man with a family dependent upon him to
a like risk, unless operation were really necessary,
or at all events very strongly called for.

These considerations have, of course, no more
exclusive or special application to the radical cure
of hernia than they have to many other surgical
proceedings. My object in drawing attention to
them is merely to insist on the fact that no definite
law can be laid down upon this question, which is
applicable with equal force to all patients. Every
case has, in truth—and this is a point which cannot
be over-emphasised—to be considered entirely upon
its own merits, not only with reference to the hernia

itself, but also with regard to many collateral matters besides.

Reserving the methods of operating for discussion in the next lecture, the case for the radical cure may be summed up as follows :—

Having regard to the necessary uncertainty and inconvenience connected with the use of trusses, and bearing in mind the undoubted fact that any person who has been the subject óf rupture in infancy or childhood is especially prone to develop hernia later in life (although the original rupture may have apparently undergone spontaneous cure after the prolonged use of instrumental pressure), the operation for the radical cure may be conscientiously recommended and practised upon sound surgical principles in the following conditions occurring in otherwise healthy people :—

(*a*) Cases of hernia in young growing subjects.

(*b*) Herniæ in which trusses are entirely useless, partially effective, or effective only at the cost of pressure which is sufficient to cause pain or serious discomfort.

(*c*) Irreducible ruptures.

(*d*) Cases in which the occupation or necessary amusements of the patient tend especially to strains of a kind likely to force the rupture down.

(*e*) Ruptures in which the use of trusses, although not actually painful, is irksome or intolerably inconvenient.

(*f*) Hernia in candidates who have been rejected, or have reason to anticipate rejection, for the public services on account of 'physical defect.'

(*g*) All cases of strangulated hernia, occurring at any period of life, in which the contents of the sac are sufficiently sound to allow of their return with perfect safety into the abdominal cavity, the state of the patient being at the same time good enough to justify the necessary prolongation of the operation.

In the event of the existence of organic disease, or if the general health is otherwise than good, the only justification for the performance of the radical cure is the certainty that the risk continually being run by the patient in consequence of the hernia is greater than the danger which would probably be caused by the operation.

Before leaving this part of my subject, I cannot refrain from again insisting upon the view that the exact merit of the radical cure is not to be decided merely upon the question of whether the patient can or cannot dispense with the use of a truss after the operation. A large number of patients require no truss at all, some should wear the instrument as a precaution, some may have to use it of necessity.

The main question in the matter is this: taking the general mass of cases of hernia, are they or are they not susceptible of an amount of benefit from the radical operation which is altogether in excess of the risk which is incurred in obtaining the result ?

For my own part, I can answer this question emphatically in the affirmative, and it is on this account that I never hesitate to recommend the treatment in suitable persons as a rational and sound surgical proceeding.

LECTURE X

ON THE RADICAL CURE (continued)

(The description of the treatment of the sac by invagination contained in this Lecture was published in the *Lancet*, September 12, 1891)

SYNOPSIS—Introductory remarks—ESSENTIAL POINTS IN THE OPERATION FOR THE RADICAL CURE—Approximation of walls of hernial canal of secondary importance with certain exceptions, but is at same time desirable in the majority of ordinary cases excepting in infants and children—No one operation for the radical cure equally suitable for all classes of case.

THE RADICAL CURE IN INFANCY AND CHILDHOOD—Inguinal hernia— Peculiarities of the parts in these subjects—Treatment of the sac— Difficulties in separating the sac from the cord in congenital and infantile herniæ—Separation rarely, if ever, impossible—No attempt to be made to remove scrotal part of sac after obliteration —Reason for this—Treatment of rings and canal—Obliteration of canal by suture as a rule unnecessary, in consequence of tendency to spontaneous closure under proper conditions—Object of closure by suture when necessary—Temporary (absorbable) sutures only required—Permanent sutures sometimes causing irritation at remote periods—Reference to case in corroboration—Method of introducing suture—Caution as to deceptive swelling of scrotum immediately after operation—Objection to use of drainage-tube in these young subjects.

Umbilical hernia—Radical cure the exception in consequence of strong tendency to natural recovery—Advantage of radical cure—Treatment of the sac and ring—Incidental reference to cure of ventral hernia.

THE RADICAL CURE IN YOUTH, MIDDLE LIFE, AND ADVANCED AGE— Remarks on gradual diminution of tendency to spontaneous cure both in congenital and acquired herniæ as age advances—Complete inguinal hernia—Treatment of the sac—By traction—Simple ligature and division—Cases suitable—Cases requiring more elaborate methods of two principal classes—(*a*) Those containing bowel or omentum only—(*b*) Those containing both bowel and omentum —Treatment by invagination of sac recommended in Class *a*— Description of author's plan—Advantages offered by it.

THE numerous and diverse methods which have been devised for the radical cure of hernia are so well dealt with in the various text-books that any systematic description of them in a course of lectures like this would be clearly superfluous, if not entirely out of place.

I shall therefore attempt to do no more than indicate the proper lines upon which it appears to me that the radical cure should be conducted, describing in detail those operations only which I have found after impartial trial to be best adapted in a general way for attaining the objects in view.

ESSENTIAL POINTS IN THE OPERATION FOR THE RADICAL CURE

All methods devised for the radical cure must, if they are to be thoroughly effectual, meet the following requirements :—

(*a*) The complete obliteration of the sac.

(*b*) The provision of an efficient barrier across the abdominal aspect of the aperture through which the rupture escapes.

(*c*) The abolition of any depression (hernial fossa) in the peritoneum over the abdominal orifice of the hernial canal.

(*d*) The exclusion of any part of the sac, whether twisted or not, from between the margins of the abdominal ring concerned.

The approximation of the pillars or margins of the
ring after the sac has been dealt with is in itself of
very secondary importance, excepting in the case of
umbilical and ventral hernia, in which the union of
the ring margins is a vital part of the operation.

In ordinary cases of inguinal rupture, the closure
of the ring by suture is generally desirable, as an
accessory measure ; in some of these cases it is, however, impossible to effect the closure in any useful
manner, and in infants and very young children it is
frequently quite unnecessary.

Before passing on to the discussion of the best
means for meeting the requirements just enume-
rated, it is in the first place needful to realise the
important primary fact that no one operation or
method yet devised can be applied with equal
efficacy in every instance of any variety of rupture.
In other words, it must be clearly understood that all
cases submitted to the radical cure should be dealt
with according to their individual necessities, which
will often vary with the age of the patient, the seat
of rupture, the shape of the sac, the suppleness
of the parts, and other minor details. It therefore
follows that a surgeon who persistently performs
in all cases of any given variety of hernia the same
operation will in the long run have far less successful results than one who is prepared to vary
the methods employed in order to better meet the

requirements of the cases coming under treatment. It may, for example, be sometimes found in operating upon a patient who has an inguinal hernia on each side that a method which is quite effective on one side may be insufficient on the other.

In describing the plans which I most frequently adopt in my own practice, I wish it therefore to be thoroughly understood that I do not for one moment intend to imply that they are intrinsically better than the operations practised by other surgeons, nor that they should in any way supersede them. Indeed, for reasons which I hope I have made clear, it is my custom to perform the method which seems best suited to the requirements of individual cases.

The only operations to which I altogether take exception are those in which some part of the sac is left between the pillars of the ring; such methods I do not now use because they appear to me to be devised on a faulty basis, and from experience I have come to see that they are also defective in result.

The condition of the structures in infancy and early childhood differ so much from that of the same parts in late youth and adult life, that it is convenient to consider the two classes of patient separately.

N

The Radical Cure in Infancy and Childhood

The peculiarities of the parts concerned in the radical cure at these periods of life are principally —(*a*) the extreme elasticity of the peritoneum and the looseness of its connection with the abdominal parietes, especially in the region of the groin, as compared with what is met with in later life; and (*b*) the natural tendency, already explained in the last lecture, to spontaneous closure of the inguinal canal, provided that no abnormal structure exists between its walls to prevent their coming together.

The radical cure under these circumstances of inguinal hernia, which is the only form of rupture in the groin met with in these young subjects, is as a rule a comparatively simple proceeding, and may be carried out with great certainty of permanent success in the following manner, which will be found applicable in nearly every case.

Treatment of the Sac.—The portion of the sac lying immediately below the external abdominal ring, together with the ring itself, should be exposed through an incision, over the long axis of the hernia, of such a length as the size of the rupture seems to indicate. In exposing the ring itself care should be taken that none of the structures forming any part of the wall of the inguinal canal (*e.g.* the inter-columnar fibres) should be divided; to this point I attach great importance. After exposure, the sac just below the external ring is carefully separated from the constituents of the spermatic cord, the separation being carried as far up the canal as possible without any division, laceration, or undue stretching of its walls.

If gentle but steady traction be now made upon the sac (the hernia, of course, being reduced), it will be found, in consequence of the looseness of the peritoneal connections just inside the belly, that its neck can be drawn quite out through the external ring. The neck having in this way been drawn down as far as seems safe, a ligature of silk, or better, fine tendon or strong catgut, is passed around it and tied as high up in the canal as practicable, whilst the sac is still being drawn down. The sac is then divided just below the ligature, and the stump springs back into the belly of its own accord.

If sufficient traction has been made, it will now be found, on introducing the finger through the canal, that the ligatured stump lies inside the parietes considerably above the margin of the inner ring, so that stretching across the orifice of the ring itself is a smooth surface of peritoneum which presents no depression (*hernial fossa*) towards the canal. Should the traction have been insufficient, the stump will be felt drawn up inside the parietes, opposite and a little distant from the upper aspect of the internal ring.

This latter position of the stump, although not so good in itself, does not invalidate the prospect of a perfect result if the margins of the canal are approximated subsequently.

As the ruptures submitted to the radical cure in infancy and childhood are generally either of the congenital or infantile variety, considerable difficulty may arise in separating the sac from the spermatic cord. All attempts at this separation should be limited to the upper part of the cord, as the connections between the vas deferens and the sac are much less intimate just inside and below the inguinal canal than they are farther down towards the scrotum. Bearing this point in mind, I have never failed to isolate the upper part of the sac in these cases.

It has been stated by some surgeons that it is occasionally impracticable to isolate the sac in consequence of the vas deferens being so completely buried in a reflection of the serous membrane that its separation is impossible. Should this rare condition of affairs be met with, it may be necessary to leave a strip of the sac attached to the vas, but every endeavour should be made to avoid this, as the thoroughness of the cure may possibly be somewhat impaired by it.

After the division of the sac no attempt should be made to remove its scrotal part, for it is in this attempt that the integrity of the testicle is endangered—indeed, it is quite easy, after having with much trouble separated the vas deferens above, to divide it below in endeavouring to get away as much of the scrotal sac as possible.

It must be borne in mind that removal of this scrotal part of the sac in congenital cases is always impossible, and is often impracticable in the infantile kind. What is left of the sac will shrivel of its own accord and give no trouble.

TREATMENT OF THE RING.—If, after having been treated in the manner described, the stump of the sac, when it has retracted into the belly, lies well away from the actual orifice of the canal, as it usually does, no approximation of the margins of the rings or canal by sutures is necessary. This I say advisedly, because some of the best results which have followed in my own practice have occurred in cases in which the parts have closed spontaneously, no sutures of any kind having been used.

If, on the other hand, after its retraction the stump lies over or close upon the abdominal orifice of the canal, and more especially if the retraction has been so slight as to leave the stump in the canal itself, then the margins should certainly be closed by sutures of some absorbable kind, such as tendon. The object of the sutures under these circumstances is merely to prevent the ligatured stump from being forced down, by coughing or otherwise, between the sides of the canal so as to present an obstacle to its spontaneous closure. They therefore need not be of a material like silk, which remains unchanged in the tissues, as they are only required temporarily during the earlier process of consolidation of the parts, and not as a permanent uniting medium.

Indeed, in some of these young subjects, sutures of silk, silkworm gut, or silver are not only unnecessary, but, by reason of their permanency, actually harmful, in consequence of the irritation which they set up.

For instance, in a case of mine, the progress of which after operation was everything that could be desired, I had to remove about a year later a silk suture used in approximating the pillars of the ring, in consequence of the irritation caused by it. Attention was called to the trouble by the child's constant general peevishness, and a continual habit which he had of clutching at the parts about the operation scar. On cutting down upon the suture it was perfectly quiet and encysted; I, however, removed it, and immediately all the symptoms complained of subsided, and nothing further was heard of them.

In this case I imagine the cause of the irritation was probably the inclusion of a nerve in the suture, an occurrence impossible to prevent always with certainty.

With regard to the method of introducing these sutures, there is nothing much to be said. They should invariably include the conjoined tendon and the whole thickness of the margins of the canal, and on the outer side, one at least should pass beneath or through the extreme edge of Poupart's ligament. Whether they are of the laced, continuous, or interrupted kind matters not the least; if interrupted, two will always be sufficient, and frequently one only is necessary.

The way of treating the wound in the superficial parts requires no special notice; it will vary with the fancy of the operator. I use no drainage-tube, close the wound with horsehair, covering it afterwards with a wet dressing of 'double cyanide' gauze.

In the early periods of life there is very prone to occur, as sometimes also happens in adults, a great deal of swelling from effusion, more particularly in congenital cases, into the scrotal part of the sac.

This is a point to be noted, as inexperienced practitioners are occasionally led to think the swelling to be due to the descent of the hernia again; indeed, I know of instances in which the scrotum has been opened up unnecessarily on this account, and it is not long since that I was asked to see a case upon which I had operated, because it was feared that the rupture had come down again, as the swelling looked so much like it.

This effusion always, so far as I have seen, subsides in a few days, the parts resuming the normal aspect. The use of a drainage-tube during the first twenty-four hours after the operation obviates this symptom; but drainage is better dispensed with in very young subjects, especially infants, whose dressings are liable to become soiled with urine, as the drain provides a channel along which peccant material may easily pass into the depths of the wound; whereas, if the edges are closely stitched and no drainage employed, healing is so quick, that if a little septic or irritating material does flow over the parts no harm results.

In *Umbilical Hernia*, which is by far the commonest form of rupture met with in the early periods of life, the tendency to spontaneous 'cure' is so strong that the question of the radical treatment rarely arises. It would, however, I believe, be distinctly advantageous if more of these cases, in which the herniæ are large, with long sacs and gaping rings, were submitted to the radical cure with the object of preventing the formation of rupture at the navel in older life, which so frequently happens in such cases, especially in women.

In the event of the radical operation being thought desirable, it should, as there is no peculiarity in the parts concerned which need be considered, be conducted upon the plan which is used in simple cases in the adult—*i.e.* the sac must be exposed, the redundant portion removed, and the margins of the ring brought together with the superficial parts in the way which is commonly employed for closing the wound in an abdominal section.

This method is also well suited for the radical cure of *ventral hernia* in these young subjects. I have myself seen only one suitable case of this kind, and that occurred in a small boy who had been struck on the abdomen with the corner of a slate by a school-fellow. A rent had apparently been made through the abdominal muscles about midway between the anterior superior iliac spine and the umbilicus, the skin having remained intact. When I saw the patient eighteen months after the injury, there was a large ventral rupture which could clearly have been successfully treated by the plan just indicated; the boy, however, was not very robust, and the parents thought the slight risk involved by the operation unjustifiable.

The Radical Cure in Youth, Middle Age, and Advanced Life

In considering the radical treatment at these periods of life, the following points already referred to in a previous lecture must not be lost sight of.

In inguinal rupture of the congenital or infantile kind (*i.e.* hernia into the funicular process), and in ordinary cases of umbilical hernia, the tendency to spontaneous cure becomes after puberty gradually less and less, until at the age of five-and-twenty it may be considered to have entirely ceased.

In acquired hernia, especially if it occurs suddenly in the *complete* form, there is at first, but only for a very short time after the original formation of the rupture, a considerable tendency to spontaneous cure in all fairly robust subjects up to the age of thirty-five, or possibly forty, if the hernia is not allowed to descend again after it has once been reduced.

After this time of life, there is little inclination to a natural cure in acquired cases.

In femoral hernia, which is essentially the hernia of middle life, no tendency to spontaneous cure exists, in consequence of the peculiarity of its production, and the character of the canal through which it comes.

Inguinal Hernia. Treatment of the Sac.—In

growing subjects, when the hernia is small, ligature of the sac in the manner described as suitable in infancy and childhood is generally sufficient to obtain a perfect result, approximation of the walls of the canal being necessary only under the conditions already indicated. If, on the other hand, the hernia is large, frequently in the sac, possibly adherent, and especially if it is increasing in size, the rings being at the same time wide and gaping, the treatment in growing patients must be conducted upon the same lines as in later life, because the tendency to a natural cure cannot be relied upon for any effectual help.

Under these circumstances, as well as in middle life and advancing age, cases of hernia regarded solely from the point of view of the *technique* of the operation for the radical cure, consist practically of two principal kinds.

In the first of these, the sac contains either bowel or omentum only, and in the second kind it is occupied by omentum, together with bowel. Cases in which the sac contains bowel only are most uniformly met with in young subjects, in whom omentum is not often found alone in the sac, and does not generally form a portion of the hernial tumour at all, excepting in unusually large ruptures which have been much neglected. In middle-aged and old subjects, on the other hand, omentum commonly forms either a considerable portion of, or, as not infrequently happens, the whole contents of the sac.

It may be accepted without reserve that when both bowel and omentum together occupy the sac, the ring and canal are larger, and the margins more rigid and more difficult to approximate, than in cases in which either bowel or omentum exists alone. It follows, therefore, as a natural consequence that cases in which the sac contains gut and omentum together offer, in an ordinary way, more obstacles to effecting a permanent ' cure' than the other kind of hernia formed of either gut or omentum alone.

CASES IN WHICH THE SAC CONTAINS BOWEL OR OMENTUM ALONE.—In cases of inguinal hernia in which the sac is occupied by a knuckle of gut only or a mass of omentum, the indications for the radical cure appear to be most effectually met, in the majority of patients, at the time of life now being considered, by simple invagination of the whole or a portion of the sac (that is to say, turning it outside in), and attaching the fundus of the invaginated part at a point some distance above the abdominal orifice of the canal through which the hernia descends.

Details of the Operation by Invagination in Inguinal Hernia.—The upper part of the sac, the external abdominal ring, and the aponeurosis around for an inch or thereabouts, are exposed freely by an incision through the superficial parts in the long axis of the hernial tumour. The sac, after having been carefully isolated from its connections just below the external ring (the lower part being left entirely undisturbed), is opened, and the contents reduced into the abdomen if they have not already returned spontaneously. Should the hernia consist of omentum only, this must be freed from adhesions, if any exist, the redundant portion ligatured and removed, and the stump returned. The sac is now divided just below the external ring, the distal portion being allowed, after all bleeding has been stopped, to drop back into the scrotum. The proximal part of the sac is next separated from the sides of the canal as high up as the internal ring by gentle manipulation. One finger (or more if the ring is large) of the left hand having been introduced into the abdominal cavity through the neck of the sac, any bowel lying near the internal ring is pressed back out of the way. An ordinary pile needle on a handle (unthreaded) is then made to enter the abdominal aponeuroses about three-quarters of an inch above the upper margin of the external ring—a little to the outer side of its middle line—and transfixes the whole of the aponeuroses and peritoneum, impinging on the end of the finger which occupies the neck of the sac. The needle, guided by the finger, is passed down the inside of the sac and made to pierce its outer wall at a point about half an inch

from the cut edge. The needle having been threaded
with a tendon or catgut suture, previously prepared,
and not less than twelve inches long, is withdrawn,
taking one end of the suture with it. The result is
that one end of the suture is seen passing into the

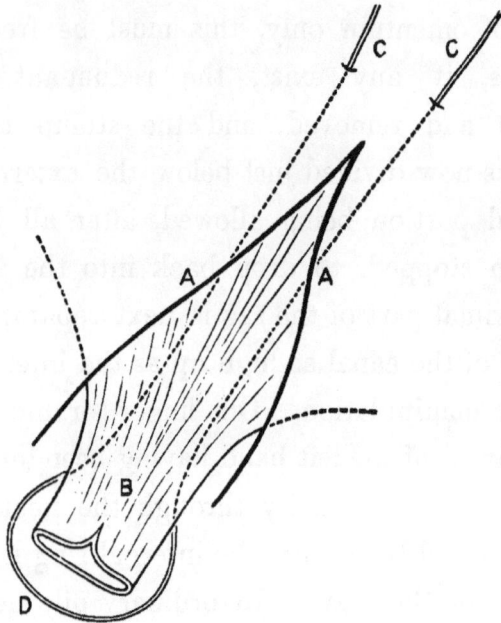

FIG. 5.

A, Pillars of ring.
B, Sac divided and isolated.
C, C, Invagination sutures passing through aponeuroses and forming loop, D.

abdominal aponeuroses above the external ring,
whilst the other issues from the outer wall of the
proximal part of the sac near its cut edge. The
needle, again unthreaded, is now made to transfix the
abdominal aponeuroses and peritoneum, about half
an inch internal to the point at which it entered
before, traversing the sac in the same way, finally

piercing the inner wall at about the same distance from the cut edge as it had done on the outer side. After having been threaded with the lower end of the suture, the needle is withdrawn, carrying the suture, as before, with it. The two ends of the sutures will now be seen entering the aponeuroses above the external

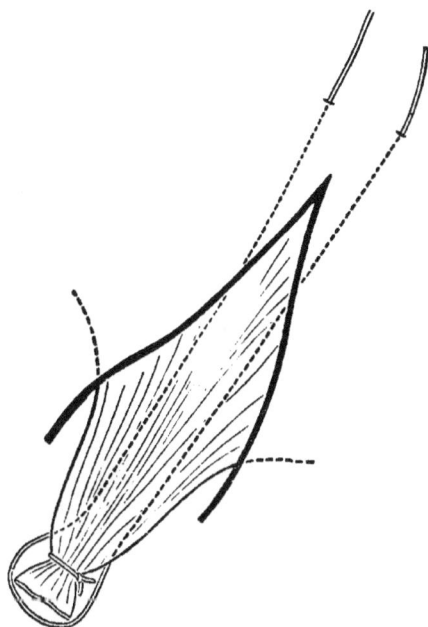

FIG. 6.—SAC OCCLUDED BY LIGATURE.

ring, and forming below a loop over the cut edge of the proximal portion of the sac, as shown in Fig. 5.

The open end of the sac is next sewn up by a continuous stitch of catgut or silk, or occluded by means of a silk ligature placed around it as close as possible to the spot at which the invagination suture pierces its side (Fig. 6). The succeeding step is the invagination of the sac, which is effected by pushing with the finger the closed end through the canal into the abdomen, the invagination sutures passing through the aponeuroses above the external ring, being at the same time drawn tight.

By this proceeding the sac is turned completely

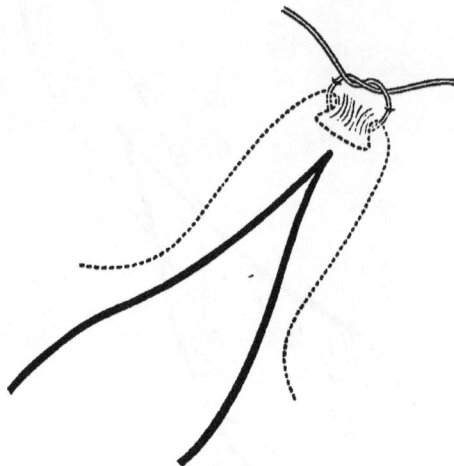

FIG. 7.

The dotted line shows the sac reduced and turned outside in.

outside in (Fig 7), and its fundus firmly attached to the peritoneal surface of the anterior abdominal wall some distance above the internal ring. The margins of the canal are then brought together in the usual way by stout silk sutures which pass through the whole thickness of each wall, a drainage-tube placed in the portion of the sac remaining in the scrotum, the superficial parts brought together with horsehair stitches, and an antiseptic dressing carefully applied.

If the hernial canal is of any considerable size, it is better, although not absolutely necessary, that

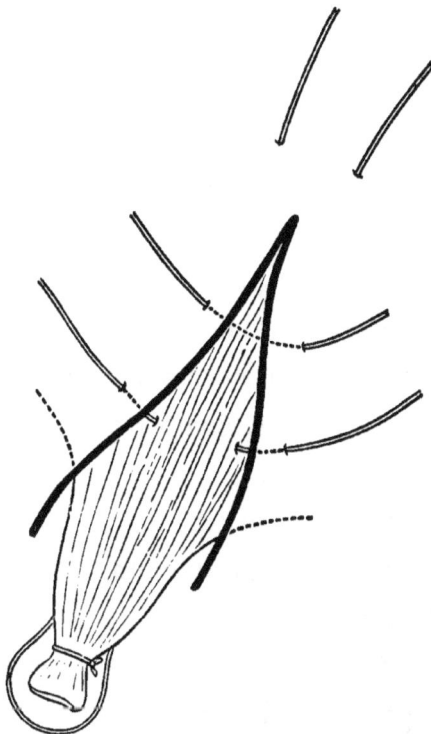

FIG. 8.

Showing sutures for approximating the sides of the canal piercing anterior wall of sac, and left loose until after the sac has been invaginated.

the upper one or two sutures used for bringing the walls together should transfix the sac (Fig. 8), passing immediately behind its anterior wall, in front of the invagination suture. This is best done before the open end of the sac is closed, in order that

o

the finger may be introduced and used as a guide to the needle as it crosses the interior of the sac, and also to protect any bowel which may by chance be forced down—*e.g.* by coughing.   The sutures which transfix the sac are left loose until the invagination has been completed, and are then tightened up, bringing the sides of the canal together, and at the same time fixing the anterior wall of the sac across the abdominal aspect of the internal ring.[1]

This plan of treating the sac, although rather complicated for description, is perfectly simple in practice, and provides the essentials for the radical cure—*i.e.* (1) a barrier across the abdominal orifice of the canal, and (2) the complete obliteration of the hernial fossa in the peritoneum by the dragging up of the posterior wall of the sac, which brings an entirely new surface of peritoneum into relation with the internal ring.

It is important to note that the barrier above the ring is composed of layers of peritoneum, placed with their serous surfaces in contact; speedy adhesion and matting together of the parts therefore follow, a much stronger barrier being thus produced

---

[1] Exception has been taken to the use of these transverse fixation sutures on the ground that they prevent the complete inversion of the sac.   A fair trial of the method will, I believe, show that this objection is practically valueless.   Further, these transverse sutures are in some cases useful in allowing the operator to determine the exact level at which the folding over of the anterior wall of the sac shall take place.

than would appear likely at first sight. Moreover, in consequence of the rapidity with which this adhesion occurs, the parts are earlier in a condition to bear pressure than is the case where the consolidation depends upon the relatively tardy union of connective tissue, by which it is effected in some of the other recognised operations for the radical cure: it is also clear that by this method no projecting mass is left over the internal ring, as the small lump formed by the fundus of the invaginated part of the sac is, if properly arranged, well away from the orifice of the canal.

The only point in the details of the proceeding which presents any difficulty is that which is common to all of these operations—viz. the separation of the sac from the constituents of the cord in cases of congenital hernia. In very rare cases this may perhaps be hardly practicable with due regard to the safety of the vas deferens, and if this should be so it is better to leave a strip of the sac attached to the cord. If, however, the attempt at isolation be confined to the upper part of the sac, it is quite exceptional for any serious difficulty to arise, for it is a good practical point to bear in mind that the nearer the sac approaches the abdominal cavity the less intimate becomes its connection with the cord. Hence, as I have already stated in speaking of the radical cure in childhood, no attempt should be made to separate the cord from the sac at any other point than just outside the external abdominal ring, the lower part being left entirely undisturbed. The isolation or removal of the lower part of the sac is always unnecessary, frequently impracticable, and, if persisted in, may endanger the integrity of the testicle, whilst it certainly prolongs the operation and serves no useful end.

My first experience of the method described was in a case of strangulated inguinal hernia about three and a half years ago; but so far as any published record is concerned Mr. Stanmore Bishop was the first, I believe, to call attention to the radical cure by complete invagination of the sac by an open operation in the 'Lancet' of March 31, 1890, where he describes a well-considered method, in which the sac is turned completely inside out and fixed across the mouth of the inguinal canal by means of a purse-string-like arrangement of sutures, in such a way that the wrinkling up of the sac produces a 'boss exactly over the internal ring.' The main advantage claimed for this operation was the rapid and firm consolidation which resulted from the apposition of the serous surfaces and the production of the boss just mentioned. It will be evident at once that my plan is identical in principle with Mr. Bishop's, but in detail there is a difference which, although at first sight trivial, is, in my opinion, important. The difference I refer to lies in the fact that, whereas one of Mr. Bishop's objects is to provide this boss over the internal ring, my intention is to avoid this by attaching the fundus of the invaginated sac well. away from the ring, in order to have a perfectly smooth peritoneal surface over the abdominal end of the canal, thus producing, in fact, as nearly as possible the condition of parts which exists in a subject who has never been ruptured at all. This, I venture to think, is better than leaving anything like a rounded lump over the internal ring, for, however this projecting mass may be arranged, it must necessarily under certain circumstances of pressure tend to act as a wedge, especially if the apposition of the sides of the inguinal canal be imperfect. I therefore believe that the smooth barrier formed by the operation I have described across the internal ring is preferable to the boss produced in the other method.

LECTURE XI

ON THE RADICAL CURE (concluded)

(The description given in this Lecture of the method of using an omental pad in the radical cure is from a paper of the Author's, published in the *Lancet* of September 19, 1891)

SYNOPSIS—INGUINAL HERNIA—TREATMENT OF SAC (CONTINUED)—CASES IN WHICH THE CONTENTS OF THE SAC CONSIST OF BOTH OMENTUM AND BOWEL—Treatment more difficult from condition of the parts —Treatment by invagination and omental graft—Description of the method—Objects attained by it—Manner in which it differs from methods in which omentum is fixed between the pillars of the ring—Objection to the method—TREATMENT OF THE RINGS AND CANAL—No closure required in certain cases if sac has been properly dealt with—Closure desirable in majority of cases—Method employed immaterial so long as certain requirements are met— Variety of suture—Importance, in author's opinion, of leaving wall of canal and rings intact—Reason for this—Whole success of cure ultimately dependent on proper management of sac.

INCOMPLETE INGUINAL HERNIA—Not well adapted for radical treatment unless nearly complete—Description of operation recommended by author in slight cases—Treatment when the rupture is nearly complete—Description of operation used by author.

FEMORAL HERNIA—Radical cure less frequently called for than in inguinal cases—Reasons—TREATMENT OF THE SAC—By simple ligature and by invagination—TREATMENT OF RING AND CANAL— No effectual method yet devised—Some of those practised quite useless—Reasons—Only effectual plan author is acquainted with too serious to be justifiable—General conclusions.

UMBILICAL HERNIA—General considerations—Description of method recommended for the radical cure in the different cases—Complication sometimes arising from difficulty in reducing large herniæ generally, and especially in umbilical rupture—Plan of treatment by the use of the omental pad—Description of operation—Cases suitable—The approximation of the margin of the ring by the flap-splitting plan.

APPENDIX—Note on the indications for the use of trusses after the radical cure.

Complete Inguinal Hernia—Treatment of Sac (continued)—Cases in which the Contents of the Sac consist of both Omentum and Bowel.—As has already been stated, the prospects of a radical cure in the true sense—*i.e.* leaving the patient entirely independent, under ordinary circumstances, of the use of trusses—are considerably less in these cases than in those previously discussed; the main obstacles to complete success being the large size of the ring and the rigidity of the tissues, which render the approximation of the walls of the canal difficult and sometimes impossible. To these must be added the entire absence of any inclination to spontaneous contraction of the parts, excepting in hernia of very recent origin, rarely seen in this class of case, which is almost always of long standing. It necessarily follows that the amount of relief ultimately obtainable will often depend solely upon the actual strength of the barrier across the internal ring, since the help afforded by the approximated walls of the canal is in most cases not of much *permanent* value. The degree of the rigidity of the parts is determined to a great extent by the chronicity of the hernia; it also appears to be in direct proportion to the amount of omentum which the sac contains, whilst it is comparatively independent of the quantity of bowel in the rupture.

It has long been an accepted principle in operations for the radical cure that all redundant or altered omentum should be ligatured and removed, the stumps being returned into the abdomen and left to their own resources. Personally, I have for some time felt that by this wholesale removal valuable material is sacrificed, some of which should be utilised for strengthening the barrier across the internal ring—certainly in all inguinal cases of this kind, and in some umbilical herniæ. Acting upon this view, I have been led to utilise the omentum in appropriate cases in the way described in the following operation, which is not intended to apply to instances in which merely a tag of recent omentum is found in the sac whilst operating upon younger subjects, in whom the ring is nearly always of moderate size, with supple margins.

Details of the operation for the radical cure by the use of an omental pad in inguinal hernia.—The parts having been exposed in the manner described in the last Lecture, the sac is opened and the bowel returned, any adhesions having been dealt with in the ordinary way. The omentum is now examined, and a flattened piece selected large enough to cover the internal ring and overlap its edges for some distance. Should there be no single piece of the desired size or shape, two smaller portions may be stitched together with carbolic catgut. In most cases, however, a suitable flap is obtainable by carefully disentangling the omental masses, especially when they have become large and nodular from spontaneous growth in the sac. It is better that the flap should be firm or even indurated, than too thin and recent. The portions of omentum not required having been ligatured and removed, two sutures of carbolised catgut about twelve inches long are passed by means of a needle through the omental flap, one on each side, a short distance from its edges, care being taken that no vessel is transfixed.

One or more fingers of the left hand are now introduced by the side of the omentum (which is not yet reduced) into the abdomen through the neck of the sac, and a pile needle on a handle (not threaded) is passed from without inwards through the abdominal aponeuroses about half an inch from the outer margin of the external ring rather below the level of its upper end. Guided by the finger, the needle passes down the inside of the sac until the point projects through the incision made in laying it open.

The needle, having been threaded with one end of the outer of the two sutures which transfix the omental flap, is withdrawn, carrying with it the suture end, which is left projecting from the abdominal aponeuroses at the point where the needle entered. The needle (again unthreaded) is now made to pierce the aponeuroses and peritoneum about half an inch below the first puncture, traverses the interior of the sac in the same way, and is then threaded with the other end of the outer suture, which is carried through the aponeuroses by the withdrawal of the needle in the same way as before. This proceeding is repeated, step by step, on the inner side of the ring, with the suture transfixing the omental flap near its inner

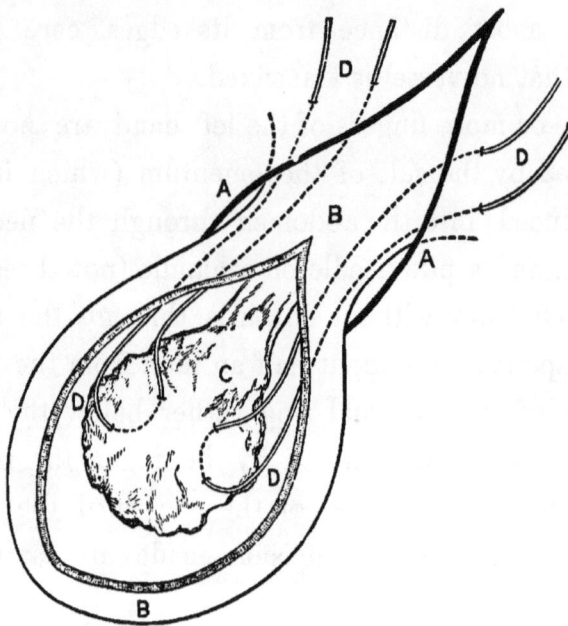

Fig. 9.

A, Pillars of ring.   B, Sac laid open.   C, Omental pad.
D, Sutures passing through abdominal aponeuroses and peritoneum above, and piercing edges of omental pad below.

edge. The resulting arrangement of sutures, &c., is seen in Fig. 9.

The omental flap, together with any ligatured stumps not previously reduced, is now carefully returned into the abdomen, the suture ends (D, Fig. 9) being at the same time drawn upon so as to bring the flap over the internal ring and retain it in position (Fig. 10). The sac is then dealt with in any manner which seems suitable. If the invagination method is used, the final tightening of the omental sutures is left until after the sac has been invagi-

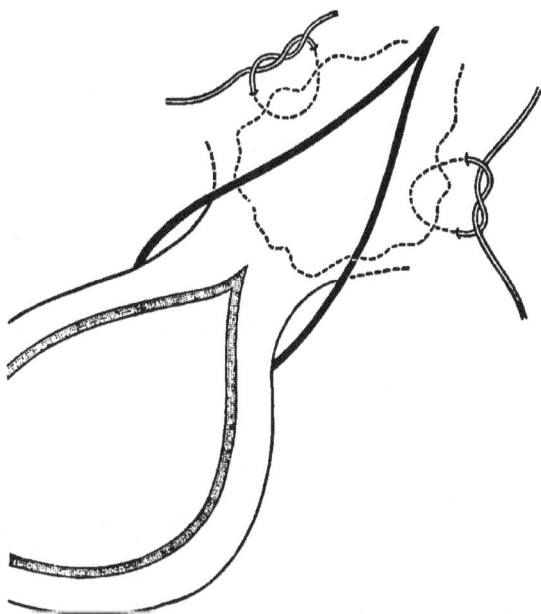

FIG. 10.

The dotted lines show the omental pad reduced and fixed in position.

nated, care being taken that the invagination suture passes down through the sac in front of the omental pad, so that when reduced and turned outside in the sac lies in front of the omental pad, between it and the peritoneum of the anterior abdominal wall. The sac having been fixed, the sutures retaining the omental pad in position are drawn tight and fastened off. The operation is completed by the bringing together of the wound in the superficial parts after the walls of the canal have been completely approximated, or as nearly so as practicable, by means of deep sutures of stout silk or fishing gut, the upper one or two of which may with advantage be made to transfix the sac in the way previously mentioned.

By this operation a barrier is formed across the internal ring, which is composed of the inverted sac in front and the omental pad behind, arranged in such a way that the serous surface of the front wall of the sac is in contact with the peritoneum on the anterior abdominal wall, whilst the omental pad is firmly fixed across the serous surface of the posterior wall of the sac. The close and firm adhesion which rapidly takes place between these parts produces a barrier of more than ordinary strength. In addition to this, the general matting together which occurs greatly helps in keeping together the walls of the canal; so that not only is a barrier provided, but also a uniting medium. In bringing together the walls of the canal it is important to use permanent sutures; silkworm gut is probably the best material, but ordinary stout silk is rather more manageable, as it can be tied more easily.

From my experience of this operation, I believe

it offers a surer prospect of complete success in the class of case for which I employ it than any other method. It is hardly needful to point out that it entirely differs from the practice which has been adopted by some operators of fixing the omentum *between the walls of the canal*, which is obviously objectionable on account of the wedge-like action which must necessarily be exercised by any plug of tissue left in the canal.

It has been said also that omentum fixed in the canal is prone to act as a guide for conducting the bowel to a weak point in the abdominal wall, but of this fact I am not at all sure.

In the operation here advocated the first of these objections, of course, does not exist, and, so far as the second is concerned, I venture to think that even if the bowel were guided by the stalk of the omental pad to the site of the original hernial orifice, the wall is there so strengthened that it would probably withstand pressure as well as the parts in the same locality would do in any subject without hernia, whose rings were not unusually strong.

One other objection remains to be noticed—viz. the possibility of intestinal obstruction resulting from a knuckle of gut becoming strangulated by the stalk of the attached piece of omentum. I do not think that much weight need be attached to this point, for although Mr. Greig Smith has mentioned a case in which strangulation was produced in this way in a patient in whom the omentum had been fixed in the ring, it must be an occurrence of such rarity as to hardly afford any valid objection to a proceeding such as that I am now advocating, in which the advantage derived from the attachment of the omentum across the internal ring altogether exceeds the risk of harm resulting from it.

Again, the fact must not be overlooked that omental stumps, after having been ligatured and returned in orthodox fashion (*i.e.* when left to their own resources), occasionally contract strange adhesions. For example, I have seen an instance in which a band, produced by the adhesion of an omental stump, was found stretching across a knuckle of the ileum, and although no obstruction had actually resulted, it was plain that most of the conditions necessary for its production were present. I am therefore inclined to think that practically the risk of obstruction is really not more when the omentum is attached than when it is left to itself.

TREATMENT OF THE RINGS AND CANAL.—I have already pointed out that if the canal is small, and when in acquired cases the rupture is of only moderate size, no suturing of the rings or canal walls is really necessary in adults up to the age of thirty or thereabouts if the sac has been efficiently dealt with.

In the majority of cases, however, it is desirable,

and in many actually necessary, to approximate the walls of the canal as an accessory measure after the obliteration of the sac.

The method employed in effecting this approximation is, so far as I can judge, of little consequence, provided—(1) that the sutures used are of a permamanent and unirritating kind; (2) that care is taken, when passing the sutures, that the whole thickness of the conjoined tendon and other tissues forming the walls of the canal is transfixed; and (3) that on the outer side one or two sutures are passed either completely under Poupart's ligament or through its extreme edge.

Personally I generally use silkworm gut, which is, I think, preferable to silk ; the latter, however, is better if the rings are very large and rigid and the tension is likely to be great, as it is more easily tied with security and its strength is more uniform.

As to the form of suture, the cross-laced stitch seems to best answer the requirements of the majority of the cases. When the canal is small, a single stitch is often all that is necessary. In cases having very large gaping rings, in which the tension during the first few days after the operation is sometimes very great, the sutures should be interrupted rather than continuous.

In dealing with the canal and rings the most important point of all, in my judgment, is to avoid incising or tearing any of the structures forming their walls or margins, for by so doing a cicatrix will be produced which is prone to stretch in spite of most careful suturing.

In expressing this view I am not forgetting that in one of the plans recently devised for the radical cure, the splitting up of the canal is an essential part of the operation, principally, it seems, for the purpose of allowing the sac to be dealt with. For my own part, I have never yet met with a case of complete inguinal hernia in which the sac could not be perfectly obliterated and securely fixed, without any interference whatever with the integrity of the walls of the canal, and I cannot therefore admit that any reason exists to justify the practice of laying open the canal either in part or completely. From other rather complicated methods by which a valve-like arrangement of the canal walls presumably results, I have been unable to obtain any special benefit.

In connection with this part of the subject another important point must be alluded to, as it is quite undeniable. *If the sac is properly obliterated very simple management only of the canal and rings is required; and, on the other hand, if the treatment of the sac is defective, no amount of ingenuity exercised upon the closing of the canal will ensure anything more than a temporary success.*

INCOMPLETE INGUINAL HERNIA (BUBONOCELE).—If

this consists only of slight bulging into the internal ring, it is, I believe, as a rule, unwise under ordinary circumstances to make any attempt at effecting the radical cure at all. Nevertheless, if by chance some reason exists which is considered sufficiently important by the patient or his responsible advisers (as may, for example, be the case with a youth who has been rejected for the army or navy on account of this affection), an operation may be undertaken with propriety, provided that those principally concerned in the matter are prepared for its possible ultimate failure.

After some experience in this class of case, I have come to the conclusion that the only rational method which affords a fair prospect of a good result is the following :—

An incision about an inch or an inch and a half in length is made parallel with Poupart's ligament over the most prominent part of the hernia, and the aponeurosis of the external oblique muscle laid bare. The internal ring is now exposed by an incision through this aponeurosis made in the direction of its fibres. The opening thus made should not extend farther in a downward direction than is absolutely necessary, as the prospect of ultimate success depends to a considerable extent upon the walls of the inguinal canal being left as intact as possible. Above, the wound through the aponeurosis passes far

P

enough to expose freely the margin of the internal ring (*i.e.* the arched fibres of the transversalis and internal oblique muscles), which must on no account be divided by the knife. The margin of the ring having been drawn upward with a retractor, in order to obtain as much room as possible without tearing the parts, the subserous areolar tissue is gently displaced with the fingers, and a fold of the peritoneum forming the bulging sac is caught up in a pair of clip forceps and drawn well out.

A small opening having been made into the ' sac ' to enable the operator to make sure that no adhesion of omentum or gut exists, the peritoneum, after having been gently separated by the fingers from its connections about the margin of the ring, is drawn out as far as possible, and a ligature passed around it which, after having been pushed up towards the belly, as far as practicable, is tied in the usual way. Any redundant peritoneum on the distal side of the ligature is then cut off and the stump allowed to retract of its own accord.

Thus anything like a bulging of the peritoneum or primitive hernial sac is completely obliterated, and a tense layer of peritoneum only remains stretched across the ring.

The subserous areolar tissue is now replaced, being retained in position, if necessary, by means of a fine continuous catgut suture, and the wound in the aponeurosis of the external oblique very carefully brought together with a continuous suture of thin silk or kangaroo tendon.

Finally, the superficial parts are united with horsehair, no drainage being used. The weak part of this proceeding, which is otherwise very neat and perfect, is the incision in the aponeurosis, which, however carefully managed, is followed by a cicatrix in the parietes, which is prone to stretch. This, however, is in itself not of so much consequence if no incision or tear is made in the muscular margin of the internal ring. Any interference with the ring margin almost invariably is followed by a yielding cicatrix, which will probably end in a bulging of intestine like that seen in a ventral hernia.

If the rupture, although incomplete, has traversed so much of the canal that it is presenting at the external ring, without having actually passed through it, the end of the sac is easily exposed by an incision, which lays bare the external ring without dividing any of the structures forming the wall of the canal. After the ring has been in this manner exposed, the sac is easily dealt with in young subjects by drawing it down and applying a ligature in the way already described; and in later life invagination of the sac can be effected as in cases of complete inguinal hernia, the margins of the canal afterwards being approximated by suture. The chances of success under these circumstances are the same as in the complete form of hernia at a similar period of life.

FEMORAL HERNIA.—The number of cases of femoral hernia in which the radical cure is required is naturally very much smaller than it is in the inguinal variety. In fact, it is quite exceptional to be called upon to perform the radical treatment in femoral rupture, unless strangulation, irreducibility, or some other condition of the parts exists, which makes herniotomy necessary or very desirable, irrespective of the question of the radical cure. The reasons for this are principally three—(1) the relative infrequency of femoral rupture ; (2) the fact that femoral hernia is essentially the kind peculiar to adult life and advanced age; all that large class of cases, therefore, occurring before and immediately after puberty requiring treatment for various reasons is wanting ; and (3) the almost exclusive limitation of femoral rupture to women whose occupations, except amongst the lowest classes, are but little interfered with by the affection.

In my own practice, although I have performed many operations for the radical cure in the course of herniotomy for painful, irreducible, or strangulated femoral hernia, I have only once been called upon to adopt the treatment in a case of simple reducible hernia of this kind, and that case occurred in a male.

So far as the TREATMENT OF THE SAC is concerned, the methods already recommended for inguinal hernia are equally applicable to the femoral form. In the majority of cases in which the rupture is of moderate size, simple ligature of the sac in the manner described in discussing inguinal hernia in young subjects is all that is really necessary. As a detail in this plan, when used in femoral cases, it is useful to know that the ligature can be applied rather higher up towards the belly if the neck of the sac is twisted for a few turns, as this proceeding more effectually frees the connections of the peritoneum around the proximal surface of the femoral ring, and so allows the neck of the sac to be pulled farther down. Upon the reduction of the ligatured stump into the abdomen, care should be taken to ascertain that it is quite free and lying well away from the ring, and unless the retraction is very complete, the stump should be fixed to the parietes by sutures. If the rupture is large or very old, the sac is, I think, as a rule best treated by invaginating it, after the plan described a few pages back.

The sac and lower abdominal aponeuroses are exposed by a vertical incision. The invagination suture passes through the aponeuroses a suitable distance above Poupart's ligament, well away from the upper margin of the femoral ring; and, due allowance being made for the difference in anatomical relations, the details are, step by step, the same as in the case of inguinal hernia. The operation is, however, more simple, as there is none of the difficulty in isolating the sac which often occurs in the inguinal cases. The only point requiring very careful management is the transfixion of the abdominal wall by the needle, in consequence of the possibility of wounding one of the larger blood-vessels—*e.g.* the epigastric or obturator—notorious for their tendency to abnormality. There should be no real risk in this matter, as the parts of the vessels concerned lie immediately beneath the peritoneum, and their pulsations can be so easily felt by the finger which passes through the neck of the sac that very ordinary care will suffice to avoid injuring them.

The effectual treatment of the RING AND CANAL in the radical cure of femoral hernia is a problem which has yet to be solved.

Closure by simple approximation of the walls with sutures is from the anatomical disposition of the parts impracticable with anything like safety, and no other plan has up to the present been devised by which an obliteration likely to be permanently useful can be brought about.

Attempts at obliterating the canal in part by stitching the fascia of the neighbourhood across its lower extremity effect, so far as I have seen, no good whatever, nor have I been able to obtain any real benefit from plugging it with pads of muscle[1] fixed in position by sutures.

The truth is that no plan for dealing with the aperture through which a femoral hernia comes will be found to be permanently good which does not include the essential detail of placing a barrier across *the upper aspect* of the femoral ring. All plans hitherto invented have, so far as I know, been limited to attempts at obliterating the opening *from below*. Now there is no doubt whatever that once a hernia enters the femoral ring, its passage down the whole length of the canal is only a matter of time, and that no amount of plugging or careful obliteration of *the lower part of the canal* will be of more than very temporary benefit in preventing its progress towards a complete rupture.

[1] Since this was written, Mr. Watson Cheyne has described in the *Lancet* (November 5, 1892) a plan of plugging the crural canal by means of a flap formed of the whole thickness of a portion of the pectineus muscle dissected up from the bone. This, on the whole, although rather severe, is, perhaps, the best practical method yet suggested; but it has the inherent defect to which I have alluded— namely, that it deals with the canal from the wrong end, and cannot afford any effectual barrier to the entrance of a hernia into the femoral ring itself. In other words, it does not relieve the obliterated sac in any useful degree from the burden of providing the resistance which is to prevent the initial re-formation of the hernia. Too short a time (about three months) had elapsed since the operation in the cases described by Mr. Cheyne to allow of any sound deduction as to the permanent good likely to be derived by the patients. I am, however, fairly confident that such *permanent* benefit as accrues in the cases will be due to the resistance produced by the obliteration of the sac, and not by the plugging of the canal.

Unfortunately, the detail alluded to can only be carried out by performing an abdominal section, by means of which in the cadaver it is quite easy to dissect up and stitch a mass of muscle, &c., across the ring. In the living this operation would, I think, be too severe and difficult to justify its performance. I am therefore of opinion that, seeing the satisfactory results which follow, when the sac is properly dealt with, although ño attempt at closing the canal or ring is made, there is no useful object to be attained by practising any of the methods at present in use for closing these parts, but that the greatest possible care should be concentrated on the treatment of the sac, which must ultimately in all these cases be relied upon to provide the means for the prevention of the re-formation of the rupture.

In considering this question the fact should not be lost sight of that the class of patients (*i.e.* women) to whom femoral hernia is almost exclusively limited, are not, excepting in the case of those following certain special occupations, subjected in their ordinary work in this country to conditions likely to cause strains so great as to render really necessary in the radical cure the additional temporary support which may be derived from the closure of the hernial canal.

UMBILICAL HERNIA.—When considered in relation to the radical cure, cases of umbilical hernia may be

for practical purposes divided into two classes—1.
Cases occurring in infants and young subjects, and
also those met with in adults, when the rupture is
small or of very moderate size. Such cases may be
subjected to the radical cure with a very good prospect of perfect success, by which I mean that the
patient will be relieved of the necessity of wearing
any form of apparatus or support, excepting perhaps
a light abdominal belt. 2. Cases of large and long-
standing rupture in adults and old people, which
are generally irreducible and in which the sac often
contains large masses of omentum and many feet
of intestine. These cases are not adapted for the
radical cure in the sense the term is usually understood, and indeed rarely present themselves for
operative treatment, excepting when actually strangulated, or on the point of becoming so. When,
under these circumstances, operation is necessary,
the opportunity should be taken to effect any improvement possible in the general condition of the
rupture, by curtailing or completely obliterating the
sac and approximating the margins of the ring, by
which means a remarkable amount of comfort can
often be afforded to a patient, whilst the risks of
future strangulation are greatly diminished.

In Class I. of these patients, *i.e.* in infancy and
tender age, and in adults when the rupture is of
small or very moderate size, the radical cure is easily

effected in the majority of instances in the following way :—The sac having been exposed and laid open, any adherent gut or omentum is released and reduced into the belly. The redundant portion of the sac and other tissues is then cut away, and after the margins of the ring have been refreshed, the whole wound is brought together by means of silkworm gut sutures, which include the peritoneum and whole thickness of the abdominal wall in the way ordinarily employed in dealing with a wound after laparotomy. In the simplest cases this is all that is required, and gene- rally attains the purpose in view. When omentum in sufficient quantity exists in the sac the firmness of the cicatrix may be considerably increased by fixing it across the peritoneal aspect of the wound in the way to be immediately described.

In older subjects (Class II.), especially if the ring is, as often happens, very large, these simple measures are insufficient, the difficulties of properly dealing with the parts increasing proportionately with the rigidity of the ring margins, and with the bulk of the sac contents. In some of these cases, in fact, when the sac contains nothing but great quantities of gut, the radical cure in any useful form is impracticable. This is partly on account of the very large size of the opening in the abdominal wall, and partly because, when large portions of the abdominal contents have remained in the hernial sac (*i.e.* outside the proper abdominal cavity) for a long period, the capacity of the abdomen appears to *decrease* as the contents of the sac *increase* in quantity. The result of this is that the replacement of the hernia, if it can be effected at all, is only possible after the use of so much pressure that the strain on the tissues when the ring margins are approximated is so great that there is some danger of the stitches giving way at once, or rapidly cutting their way out before sufficient union has taken place to withstand the great pressure from within.

The tendency to the gradual diminution of the belly capacity under the circumstances alluded to is a matter of considerable interest and importance in relation to the reducibility of large herniæ generally, and it is, I think, not quite so universally recognised as it should be. In large umbilical herniæ it is particularly important, for it will be found that many of these ruptures, even when of enormous size and quite irreducible under ordinary circumstances, will gradually diminish of their own accord if the patient be kept in bed for a long time, and finally become quite reducible. The reason of this is that, in consequence of the position of the subject, the gut tends to return of its own accord into the abdomen, and by the steady, gentle expanding pressure thus spontaneously exerted on the abdominal walls, the capacity of the belly, previously unnaturally diminished, is gradually increased until it so nearly resumes its original condition that it will retain without much tension the whole of the hernia. This is a clinical point of considerable importance, not only in relation to the radical cure, but also in connection with the palliative treatment of very large herniæ of all kinds.

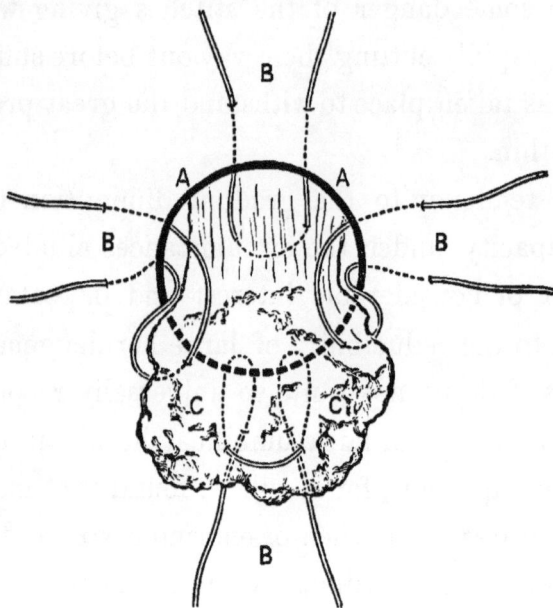

IG. 11.

A, Margin of ring.   B, Sutures.   C, Omental pad.

Umbilical hernia, even when of considerable size, may not infrequently be successfully treated by exposing the sac, cutting away its superfluous parts, and bringing together the margins of the rings by means of *permanent sutures* (silkworm gut is the *best material*, but silk is more easily applied), which include the peritoneum and whole thickness of the parietes excepting the skin, which is afterwards united over the deep stitches by horsehair. If the tension is not too great, this plan answers well. When the sac contains omentum, great help is ultimately derived if a pad or flap of the omental tissue is fixed across the abdominal aspect of the hernial orifice, in the following manner, which is similar to that already

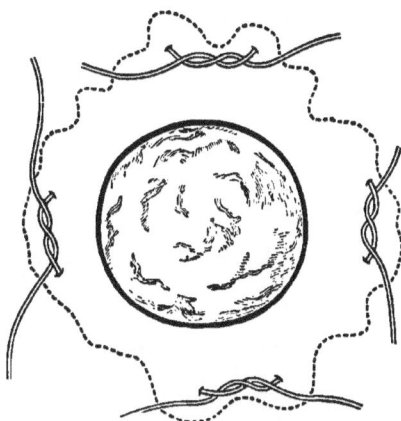

FIG. 12.

described in discussing the treatment of certain cases of inguinal hernia.

The sac is exposed and freely laid open. Any adhesions which exist having been dealt with, the bowel is reduced. The sac is then dissected out and cut as near the margin of the ring as seems proper. A suitable pad of omentum is next selected, the remainder being removed or returned into the abdomen, and four sutures, passing through the aponeuroses and peritoneum, a short distance from the margin of the umbilical opening, are arranged as shown in Fig. 11. After reduction and fixation of the omental pad across the back of the ring, in the manner indicated in Fig. 12, the margins of the ring are brought together by deep sutures of stout silk or tendon, which include the peritoneum, the operation being completed by uniting the edges of the wound in the superficial parts.

This method may also be useful in some of those cases in which the ring is large and its margins so resistent that complete approximation, if not actually impossible, can only be effected at the expense of great tension. In such cases the adhesion of the omental flap across the back of the umbilical orifice not only helps to prevent the margins of the ring from dragging apart, but also obliterates the depression along the posterior aspect of the apposed edges, which is otherwise prone to form a fossa for the lodgment of gut, which by gradual distension of the parts leads to a reproduction of the hernia, in the same way that the ventral herniæ seen too commonly after abdominal section are developed.

When the tension in any case is likely to be very considerable, either from the large size of the ring or the rigidity of its margins, there is no doubt that a firmer cicatrix results if the parts, after the redundant portion of the sac has been removed, are brought together in two distinct layers, formed by splitting up the margins into a superficial and deep part by an incision made along the edge on each side, which extends well into the corresponding rectus muscle. The deep flaps with the peritoneum are then united by permanent sutures of stout silk and the superficial ones brought together with silk-worm gut stitches, which are removed in the ordinary way when union is complete. By this 'flap-splitting' plan the deeper tissues seem to be released to a great extent, and certainly come together with much less tension than when approximation is attempted with the ri g margins unsplit.

APPENDIX TO LECTURE XI

Note on the Indications for the Use of Trusses after the Radical Cure.—On this point it is difficult to lay down any definite rule which will apply equally to all classes of hernia. In femoral and inguinal ruptures it will be found in a general way that, if the sac has been completely obliterated, no truss at all will be required in growing subjects or in healthy adults up to the age of thirty-five or there-abouts ; the completeness of the obliteration of the sac being determined by the *absence of impulse below the site of the internal ring* in inguinal hernia, and below the *level of Poupart's ligament* in femoral rupture. If abnormal impulse be felt in either case below the points indicated, it is quite certain that the sac has not been sufficiently obliterated, and that, in fact, there still remains a small pouch-like fossa in the peritoneum extending into the ring, after the manner of the primitive sac in bubonocele, which, if a truss is not used, will inevitably lead to recurrence of the rupture, no matter how ingeniously the walls of the canal have been brought together.

The same indication may safely be relied upon in the majority of middle-aged subjects. If, however, the ring is very large and patulous, as so often happens in large and long-existent ruptures, a truss should as a precaution be worn under any circumstances, unless the patient's life is one of ease and rest. Persons over forty years of age, whose occupations entail heavy manual work or great exertion, should, I think, wear trusses in all circumstances.

Exception to this may perhaps be made in cases in which the hernia operated upon is small and recent, and the patient very robust. As a rule, however, even then it is better to err on the side of safety, and recommend the employment of a truss, at all events during great exertion.

Patients with very flabby parietes, especially if the belly is inclined to be pendulous, should certainly use trusses or abdominal belts after the radical cure. This pendulous condition of parts tends, I believe, more to recurrence of the hernia after the radical treatment than any other condition (excepting, of course, the ineffectual treatment of the sac).

The comfort which an abdominal belt affords to persons with distinct tendency to rupture, or who have small bubonocele, is very remarkable; moreover, the belt also appears to exert a deterrent effect upon the progress of the affection. When the abdomen is pendulous the benefit is still more marked. In all such cases, therefore, after the radical cure a belt, made long and curving well into the groin, should be used. A hernia pad can easily be adapted to the belt if required, but the general support afforded to the flaccid abdominal walls appears to have such a good effect that the pad is rarely called for. In old people who suffer from bubonocele, as so many do, I have sometimes found the belt arranged in this way really of more benefit than a truss. This would at first sight hardly appear likely, but from experience I can certainly say that in some cases it is as I have stated.

PRINTED BY
SPOTTISWOODE AND CO., NEW-STREET SQUARE
LONDON

A LIST OF WORKS ON
MEDICINE, SURGERY

AND

GENERAL SCIENCE,

PUBLISHED BY

LONGMANS, GREEN & CO.,

39, PATERNOSTER ROW, LONDON.
15, EAST 16ᵗʰ ST., NEW YORK.

𝔐𝔢𝔡𝔦𝔠𝔞𝔩 𝔞𝔫𝔡 𝔖𝔲𝔯𝔤𝔦𝔠𝔞𝔩 𝔚𝔬𝔯𝔨𝔰.

ASHBY. NOTES ON PHYSIOLOGY FOR THE USE OF STUDENTS PREPARING FOR EXAMINATION. By HENRY ASHBY, M.D. Lond., F.R.C.P., Physician to the General Hospital for Sick Children, Manchester; formerly Demonstrator of Physiology, Liverpool School of Medicine. Fifth Edition, thoroughly revised. With 134 Illustrations. Fcap. 8vo, price 5s.

ASHBY AND WRIGHT. THE DISEASES OF CHILDREN, MEDICAL AND SURGICAL. By HENRY ASHBY, M.D. Lond., F.R.C.P., Physician to the General Hospital for Sick Children, Manchester; Lecturer and Examiner in Diseases of Children in the Victoria University; and G. A. WRIGHT, B.A., M.B. Oxon., F.R.C.S. Eng., Assistant Surgeon to the Manchester Royal Infirmary and Surgeon to the Children's Hospital. Examiner in Surgery in the University of Oxford. Enlarged and Improved Edition. With 178 Illustrations. 8vo, price 24s.

BARKER. A SHORT MANUAL OF SURGICAL OPERATIONS, HAVING SPECIAL REFERENCE TO MANY OF THE NEWER PROCEDURES. By ARTHUR E. J. BARKER, F.R.C.S., Surgeon to University College Hospital, Teacher of Practical Surgery at University College, Professor of Surgery and Pathology at the Royal College of Surgeons of England. With 61 Woodcuts in the Text. Crown 8vo, price 12s. 6d.

BENNETT.—*WORKS by WILLIAM H. BENNETT, F.R.C.S.. Surgeon to St. George's Hospital; Member of the Board of Examiners, Royal College of Surgeons of England.*

CLINICAL LECTURES ON VARICOSE VEINS OF THE LOWER EXTREMITIES. With 3 Plates. 8vo. 6s.

ON VARICOCELE: A PRACTICAL TREATISE. With 4 Tables and a Diagram. 8vo. 5s.

BENTLEY. A TEXT-BOOK OF ORGANIC MATERIA MEDICA.

Comprising a Description of the Vegetable and Animal Drugs of the British Pharmacopœia, with some others in common use. Arranged Systematically and Especially Designed for Students. By ROBERT BENTLEY, M.R.C.S. Eng., F.L.S., Fellow of King's College, London ; Honorary Member of the Pharmaceutical Society of Great Britain, &c. &c. ; one of the three Editors of the "British Pharmacopœia," 1885. With 62 Illustrations on Wood. Crown 8vo, price 7s. 6d.

COATS. A MANUAL OF PATHOLOGY. By JOSEPH COATS,

M.D., Pathologist to the Western Infirmary and the Sick Children's Hospital, Glasgow; Lecturer on Pathology in the Western Infirmary ; Examiner in Pathology in the University of Glasgow ; formerly Pathologist to the Royal Infirmary, and President of the Pathological and Clinical Society of Glasgow. Second Edition. Revised and mostly Re-written. With 364 Illustrations. 8vo, price 31s. 6d.

COOKE.—*WORKS by THOMAS COOKE, F.R.C.S. Eng., B.A., B.Sc.,*
M.D. Paris, Senior Assistant Surgeon to the Westminster Hospital, and Lecturer at the School of Anatomy, Physiology, and Surgery.

TABLETS OF ANATOMY. Being a Synopsis of Demonstrations given in the Westminster Hospital Medical School in the years 1871–75. Eighth Thousand, being a selection of the Tablets believed to be most useful to Students generally. Post 4to, price 7s. 6d.

APHORISMS IN APPLIED ANATOMY AND OPERATIVE SURGERY. Including 100 Typical *vivâ voce* Questions on Surface Marking, &c. Crown 8vo, 3s. 6d.

DISSECTION GUIDES. Aiming at Extending and Facilitating such Practical Work in Anatomy as will be specially useful in connection with an ordinary Hospital Curriculum. 8vo, 10s. 6d.

DICKINSON.—*WORKS by W. HOWSHIP DICKINSON, M.D.*
Cantab., F.R.C.P., Physician to, and Lecturer on Medicine at, St. George's Hospital ; Consulting Physician to the Hospital for Sick Children ; Corresponding Member of the Academy of Medicine of New York.

ON RENAL AND URINARY AFFECTIONS. Complete in Three Parts, 8vo, with 12 Plates and 122 Woodcuts. Price £3 4s. 6d. cloth.

⁎ The Parts can also be had separately, each complete in itself, as follows :—

PART I.—*Diabetes,* price 10s. 6d. sewed, 12s. cloth.

,, II.—*Albuminuria,* price £1 sewed, £1 1s. cloth.

,, III.—*Miscellaneous Affections of the Kidneys and Urine,* price £1 10s. sewed, £1 11s. 6d. cloth.

THE TONGUE AS AN INDICATION OF DISEASE; being the Lumleian Lectures delivered at the Royal College of Physicians in March, 1888. 8vo, price 7s. 6d.

THE HARVEIAN ORATION ON HARVEY IN ANCIENT AND MODERN MEDICINE. Crown 8vo, 2s. 6d.

ERICHSEN.—*WORKS by JOHN ERIC ERICHSEN, F.R.S., LL.D.* (*Edin.*), *Hon. M. Ch. and F.R.C.S. (Ireland), Surgeon Extraordinary to H.M. the Queen; President of University College, London; Fellow and Ex-President of the Royal College of Surgeons of England; Emeritus Professor of Surgery in University College; Consulting-Surgeon to University College Hospital, and to many other Medical Charities.*

THE SCIENCE AND ART OF SURGERY; A TREATISE ON SURGICAL INJURIES, DISEASES, AND OPERATIONS. The Ninth Edition, Edited by Professor BECK, M.S. & M.B. (Lond.), F.R.C.S., Surgeon to University College Hospital, &c. Illustrated by 1025 Engravings on Wood. 2 Vols. 8vo, price 48s.

ON CONCUSSION OF THE SPINE, NERVOUS SHOCKS, and other Obscure Injuries of the Nervous System in their Clinical and Medico-Legal Aspects. New and Revised Edition. Crown 8vo, 10s. 6d.

GAIRDNER AND COATS. ON THE DISEASES CLASSIFIED by the REGISTRAR-GENERAL as TABES MESENTERICA. LECTURES TO PRACTITIONERS. By W. T. GAIRDNER, M.D., LL.D. On the PATHOLOGY of PHTHISIS PULMONALIS. By JOSEPH COATS, M.D. With 28 Illustrations. 8vo, price 12s. 6d.

GARROD.—*WORKS by Sir ALFRED BARING GARROD, M.D., F.R.S., &c.; Physician Extraordinary to H.M. the Queen; Consulting Physician to King's College Hospital; late Vice-President of the Royal College of Physicians.*

A TREATISE ON GOUT AND RHEUMATIC GOUT (RHEUMATOID ARTHRITIS). Third Edition, thoroughly revised and enlarged; with 6 Plates, comprising 21 Figures (14 Coloured), and 27 Illustrations engraved on Wood. 8vo, price 21s.

THE ESSENTIALS OF MATERIA MEDICA AND THERAPEUTICS. The Thirteenth Edition, revised and edited, under the supervision of the Author, by NESTOR TIRARD, M.D. Lond., F.R.C.P., Professor of Materia Medica and Therapeutics in King's College, London, &c. Crown 8vo, price 12s. 6d.

GARROD. AN INTRODUCTION TO THE USE OF THE LARYNGOSCOPE. By ARCHIBALD G. GARROD, M.A., M.B. Oxon., M.R.C.P. With Illustrations. 8vo, price 3s. 6d.

GRAY. ANATOMY, DESCRIPTIVE AND SURGICAL. By HENRY GRAY, F.R.S., late Lecturer on Anatomy at St. George's Hospital. The Twelfth Edition, re-edited by T. PICKERING PICK, Surgeon to St. George's Hospital; Member of the Court of Examiners, Royal College of Surgeons of England. With 615 large Woodcut Illustrations, a large proportion of which are Coloured, the Arteries being coloured red, the Veins blue, and the Nerves yellow. The attachments of the muscles to the bones, in the section on Osteology, are also shown in coloured outline. Royal 8vo, price 36s.

HALLIBURTON. A TEXT-BOOK OF CHEMICAL PHYSIO-
LOGY AND PATHOLOGY. By W. D. HALLIBURTON, M.D.,
B.Sc., M.R.C.P., Professor of Physiology at King's College, London; Lecturer
on Physiology at the London School of Medicine for Women ; late Assistant
Professor of Physiology at University College, London. With 104 Illustra-
tions. 8vo, 28s.

HASSALL.—*WORKS by ARTHUR HILL HASSALL, M.D. London.*
SAN REMO CLIMATICALLY AND MEDICALLY CON-
SIDERED. New Edition, with 30 Illustrations. Crown 8vo, price 5s.
THE INHALATION TREATMENT OF DISEASES OF THE
ORGANS OF RESPIRATION, INCLUDING CONSUMP-
TION. With numerous Illustrations. Crown 8vo, price 12s. 6d.

HERON. EVIDENCES OF THE COMMUNICABILITY OF
CONSUMPTION. By G. A. HERON, M.D. (Glas.), F.R.C.P., Physician to the City of London Hospital for Diseases of the Chest. 8vo, 7s. 6d.

HEWITT. ON SEVERE VOMITING DURING PREGNANCY :
a Collection and Analysis of Cases, with Remarks on Treatment. By GRAILY
HEWITT, M.D. Lond., F.R.C.P., F.R.S. Ed., Emeritus Professor of
Obstetric Medicine, University College; Consulting Obstetric Physician to
University College Hospital, &c., &c. 8vo, 6s.

HOLMES. A SYSTEM OF SURGERY, Theoretical and Practical.
Edited by TIMOTHY HOLMES, M.A.; and J. W. HULKE, F.R.S.,
Surgeon to the Middlesex Hospital. Third Edition, in Three Volumes, with
Coloured Plates and numerous Illustrations. 3 Vols., royal 8vo, price £4 4s.

LITTLE. ON IN-KNEE DISTORTION (GENU VALGUM): Its
Varieties and Treatment with and without Surgical Operation. By W. J.
LITTLE, M.D., F.R.C.P.; Author of "The Deformities of the Human
Frame," &c. Assisted by MUIRHEAD LITTLE, M.R.C.S., L.R.C.P.
With 40 Woodcut Illustrations. 8vo, price 7s. 6d.

LIVEING.—*WORKS by ROBERT LIVEING, M.A. & M.D. Cantab,
F.R.C.P. Lond., &c., Physician to the Department for Diseases of the Skin at the
Middlesex Hospital, &c.*
HANDBOOK ON DISEASES OF THE SKIN. With especial
reference to Diagnosis and Treatment. Fifth Edition, revised and enlarged.
Fcap. 8vo, price 5s.
ELEPHANTIASIS GRÆCORUM, OR TRUE LEPROSY;
Being the Goulstonian Lectures for 1873. Cr. 8vo, 4s. 6d.

LONGMORE.—*WORKS by Surgeon-General Sir T. LONGMORE,* C.B., F.R.C.S., Honorary Surgeon to H.M. Queen Victoria; Professor of Military Surgery in the Army Medical School.

THE ILLUSTRATED OPTICAL MANUAL; OR, HAND-BOOK OF INSTRUCTIONS FOR THE GUIDANCE OF SURGEONS IN TESTING QUALITY AND RANGE OF VISION, AND IN DISTINGUISHING AND DEALING WITH OPTICAL DEFECTS IN GENERAL. Illustrated by 74 Drawings and Diagrams by Inspector-General Dr. MACDONALD, R.N., F.R.S., C.B. Fourth Edition. 8vo, price 14s.

GUNSHOT INJURIES. Their History, Characteristic Features, Complications, and General Treatment ; with Statistics concerning them as they are met with in Warfare. With 58 Illustrations. 8vo, price 31s. 6d.

RICHARD WISEMAN, SURGEON AND SERGEANT-SURGEON TO CHARLES II. A Biographical Study. With Portrait. 8vo. 10s. 6d.

MURCHISON.—*WORKS by CHARLES MURCHISON, M.D.,* LL.D., F.R.S., &c., Fellow of the Royal College of Physicians; late Physician and Lecturer on the Principles and Practice of Medicine, St. Thomas's Hospital.

A TREATISE ON THE CONTINUED FEVERS OF GREAT BRITAIN. Edited by W. CAYLEY, M.D., F.R.C.P. With 6 Coloured Plates and Lithographs, 19 Diagrams and 20 Woodcut Illustrations. 8vo, price 25s.

CLINICAL LECTURES ON DISEASES OF THE LIVER, JAUNDICE, AND ABDOMINAL DROPSY; Including the Croonian Lectures on Functional Derangements of the Liver, delivered at the Royal College of Physicians in 1874. Revised by T. LAUDER BRUNTON, M.D. 8vo, price 24s.

NEWMAN. ON THE DISEASES OF THE KIDNEY AMENABLE TO SURGICAL TREATMENT. Lectures to Practitioners. By DAVID NEWMAN, M.D., Surgeon to the Western Infirmary Out-Door Department ; Pathologist and Lecturer on Pathology at the Glasgow Royal Infirmary ; Examiner in Pathology in the University of Glasgow ; Vice-President Glasgow Pathological and Clinical Society. 8vo, price 16s.

OWEN. A MANUAL OF ANATOMY FOR SENIOR STUDENTS. By EDMUND OWEN, M.B., F.R.S.C., Senior Surgeon to the Hospital for Sick Children, Great Ormond Street, Surgeon to St. Mary's Hospital, London, and co-Lecturer on Surgery, late Lecturer on Anatomy in its Medical School. With 210 Illustrations. Crown 8vo, price 12s. 6d.

PAGET.—*WORKS by Sir JAMES PAGET, Bart., F.R.S., D.C.L. Oxon., LL.D. Cantab., &c., Sergeant-Surgeon to the Queen, Surgeon to the Prince of Wales, Consulting Surgeon to St. Bartholomew's Hospital.*

LECTURES ON SURGICAL PATHOLOGY, Delivered at the Royal College of Surgeons of England. Fourth Edition, re-edited by the AUTHOR and W. TURNER, M.B. 8vo, with 131 Woodcuts, price 21s.

CLINICAL LECTURES AND ESSAYS. Edited by F. HOWARD MARSH, Assistant-Surgeon to St. Bartholomew's Hospital. Second Edition, revised. 8vo, price 15s.

STUDIES OF OLD CASE-BOOKS. 8vo, 8s. 6d.

POOLE. COOKERY FOR THE DIABETIC. By W. H. and Mrs. POOLE. With Preface by Dr. PAVY. Fcap. 8vo. 2s. 6d.

QUAIN. QUAIN'S (JONES) ELEMENTS OF ANATOMY. The Tenth Edition. Edited by EDWARD ALBERT SCHÄFER, F.R.S., Professor of Physiology and Histology in University College, London; and GEORGE DANCER THANE, Professor of Anatomy in University College, London. (In three volumes.)

VOL. I., PART I. EMBRYOLOGY. By Professor SCHÄFER. Illustrated by 200 Engravings, many of which are coloured. Royal 8vo, 9s. [*Ready.*

VOL. I., PART II. GENERAL ANATOMY OR HISTOLOGY. By Professor SCHÄFER. Illustrated by nearly 500 Engravings, many of which are coloured. Royal 8vo, 12s. 6d. [*Ready.*

VOL. II., PART I. OSTEOLOGY. By Professor THANE. Illustrated by 168 Engravings. Royal 8vo, 9s. [*Ready.*

VOL. II., PART II. ARTHROLOGY, MYOLOGY, ANGEIOLOGY. By Professor THANE. Illustrated by 255 Engravings, many of which are Coloured. Royal 8vo, 18s. [*Ready.*

VOL. III., PART I. SPINAL CORD AND BRAIN. By Professor SCHÄFER. Illustrated by 139 Engravings. Royal 8vo, 12s. 6d. [*Ready.*

VOL. III., PART II. PERIPHERAL NERVES & SENSE ORGANS. [*In preparation.*

VOL. III., PART III. VISCERA. [*In preparation.*

QUAIN. A DICTIONARY OF MEDICINE; Including General Pathology, General Therapeutics, Hygiene, and the Diseases peculiar to Women and Children. By Various Writers. Edited by SIR RICHARD QUAIN, Bart., M.D., F.R.S., Physician Extraordinary to H.M. the Queen, Fellow of the Royal College of Physicians, Consulting Physician to the Hospital for Consumption, Brompton. Seventeenth Thousand; pp. 1,836, with 138 Illustrations engraved on wood. 1 Vol. medium 8vo, price 31s. 6d. cloth. To be had also in Two Volumes, price 34s. cloth.

RICHARDSON. THE ASCLEPIAD. A Book of Original Research in the Science, Art, and Literature of Medicine. By BENJAMIN WARD RICHARDSON, M.D., F.R.S. Published Quarterly, price 2s. 6d. Volumes for 1884, 1885, 1886, 1887, 1888, 1889, 1890 & 1891, 8vo, price 12s. 6d. each.

SALTER. DENTAL PATHOLOGY AND SURGERY. By S. JAMES A. SALTER, M.B., F.R.S., Examiner in Dental Surgery at the Royal College of Surgeons; Dental Surgeon to Guy's Hospital. With 133 Illustrations. 8vo, price 18s.

SCHÄFER. THE ESSENTIALS OF HISTOLOGY: Descriptive and Practical. For the Use of Students. By E. A. SCHÄFER, F.R.S., Jodrell Professor of Physiology in University College, London; Editor of the Histological Portion of Quain's "Anatomy." Illustrated by more than 300 Figures, many of which are new. Third Edition, Revised and Enlarged. 8vo, 7s. 6d. (Interleaved, 10s.)

SEMPLE. ELEMENTS OF MATERIA MEDICA AND THERAPEUTICS. Including the whole of the Remedies of the British Pharmacopœia of 1885 and its Appendix of 1890. With 440 Illustrations. By C. E. ARMAND SEMPLE, B.A., M.B. Camb., L.S.A., M.R.C.P. Lond.; Member of the Court of Examiners, and late Senior Examiner in Arts at Apothecaries' Hall, &c. Crown 8vo, price 10s. 6d.

SISLEY.—*WORKS by RICHARD SISLEY, M.D.*
EPIDEMIC INFLUENZA. Notes on its Origin and Method of Spread. Royal 8vo, 7s. 6d.

A STUDY OF INFLUENZA, and the Laws of England concerning Infectious Diseases. A Paper read before the Society of Medical Officers of Health, January 18, 1892, with Appendix of Counsel's Opinion, &c. 8vo, 3s. 6d.

SMITH (H. F.). THE HANDBOOK FOR MIDWIVES. By HENRY FLY SMITH, B.A., M.B. Oxon., M.R.C.S. Second Edition. With 41 Woodcuts. Crown 8vo, price 5s.

STEEL.—*WORKS by JOHN HENRY STEEL, F.R.C.V.S., F.Z.S., A.V.D., late Professor of Veterinary Science and Principal of Bombay Veterinary College.*
A TREATISE ON THE DISEASES OF THE DOG; being a Manual of Canine Pathology. Especially adapted for the use of Veterinary Practitioners and Students. 88 Illustrations. 8vo, 10s. 6d.

STEEL.— *WORKS by JOHN HENRY STEEL—continued.*

A TREATISE ON THE DISEASES OF THE OX; being a Manual of Bovine Pathology specially adapted for the use of Veterinary Practitioners and Students. 2 Plates and 117 Woodcuts. 8vo, 15s.

A TREATISE ON THE DISEASES OF THE SHEEP; being a Manual of Ovine Pathology for the use of Veterinary Practitioners and Students. With Coloured Plate, and 99 Woodcuts. 8vo, 12s.

"STONEHENGE." THE DOG IN HEALTH AND DISEASE. By "STONEHENGE." With 84 Wood Engravings. Square crown 8vo, 7s. 6d.

WALLER. AN INTRODUCTION TO HUMAN PHYSIOLOGY. By AUGUSTUS D. WALLER, M.D., Lecturer on Physiology at St. Mary's Hospital Medical School, London; late External Examiner at the Victorian University. With 292 Illustrations. 8vo, 18s.

WEST.— *WORKS by CHARLES WEST, M.D., &c., Founder of and formerly Physician to the Hospital for Sick Children.*

LECTURES ON THE DISEASES OF INFANCY AND CHILDHOOD. Seventh Edition. 8vo, 18s.

THE MOTHER'S MANUAL OF CHILDREN'S DISEASES. Fcp. 8vo, 2s. 6d.

WILKS AND MOXON. LECTURES ON PATHOLOGICAL ANATOMY. By SAMUEL WILKS, M.D., F.R.S., Consulting Physician to, and formerly Lecturer on Medicine and Pathology at, Guy's Hospital, and the late WALTER MOXON, M.D., F.R.C.P., Physician to, and some time Lecturer on Pathology at, Guy's Hospital. Third Edition, thoroughly Revised. By SAMUEL WILKS, M.D., LL.D., F.R.S. 8vo, 18s.

WILLIAMS. PULMONARY CONSUMPTION: ITS ETIOLOGY, PATHOLOGY, AND TREATMENT. With an Analysis of 1,000 Cases to Exemplify its Duration and Modes of Arrest. By C. J. B. WILLIAMS, M.D., LL.D., F.R.S., F.R.C.P., Senior Consulting Physician to the Hospital for Consumption, Brompton; and CHARLES THEODORE WILLIAMS, M.A., M.D. Oxon., F.R.C.P., Senior Physician to the Hospital for Consumption, Brompton. Second Edition, Enlarged and Re-written by Dr. C. THEODORE WILLIAMS. With 4 Coloured Plates and 10 Woodcuts. 8vo, 16s.

YOUATT.— *WORKS by WILLIAM YOUATT.*

THE HORSE. Revised and Enlarged by W. WATSON, M.R.C.V.S. Woodcuts. 8vo, 7s. 6d.

THE DOG. Revised and Enlarged. Woodcuts. 8vo, 6s.

General Scientific Works.

ARNOTT. THE ELEMENTS OF PHYSICS OR NATURAL PHILOSOPHY. By NEIL ARNOTT, M.D. Edited by A. BAIN, LL.D. and A. S. TAYLOR, M.D., F.R.S. Woodcuts. Crown 8vo, 12s. 6d.

BENNETT AND MURRAY. A HANDBOOK OF CRYPTOGAMIC BOTANY. By A. W. BENNETT, M.A., B.Sc., F.L.S., and GEORGE R. MILNE MURRAY, F.L.S. With 378 Illustrations. 8vo, 16s.

CLERKE. THE SYSTEM OF THE STARS. By AGNES M. CLERKE, Author of "A History of Astronomy during the Nineteenth Century." With 6 Plates and Numerous Illustrations. 8vo, 21s.

CLODD. THE STORY OF CREATION. A Plain Account of Evolution. By EDWARD CLODD, Author of "The Childhood of the World," &c. With 77 Illustrations. Crown 8vo, 3s. 6d.

CROOKES. SELECT METHODS IN CHEMICAL ANALYSIS (chiefly Inorganic). By W. CROOKES, F.R.S., V.P.C.S., Editor of "The Chemical News." Second Edition, re-written and greatly enlarged. Illustrated with 37 Woodcuts. 8vo, 24s.

CULLEY. A HANDBOOK OF PRACTICAL TELEGRAPHY. By R. S. CULLEY, M.I.C.E., late Engineer-in-Chief of Telegraphs to the Post Office. Eighth Edition, completely revised. With 135 Woodcuts and 17 Plates, 8vo, 16s.

EARL. THE ELEMENTS OF LABORATORY WORK. A Course of Natural Science for Schools. By A. G. EARL, M.A., F.C.S., late Scholar of Christ College, Cambridge; Science Master at Tonbridge School. With 57 Diagrams and numerous Exercises and Questions. Crown 8vo, 4s. 6d.

GANOT. ELEMENTARY TREATISE ON PHYSICS; Experimental and Applied, for the use of Colleges and Schools. Translated and edited from GANOT's *Eléments de Physique* (with the Author's sanction) by E. ATKINSON, Ph.D., F.C.S., Professor of Experimental Science, Staff College, Sandhurst. Twelfth Edition, revised and enlarged, with 5 Coloured Plates and 923 Woodcuts. Large crown 8vo, 15s.

GANOT.—*WORKS by GANOT continued.*

NATURAL PHILOSOPHY FOR GENERAL READERS AND YOUNG PERSONS; Being a Course of Physics divested of Mathematical Formulæ, and expressed in the language of daily life. Translated from GANOT's *Cours de Physique* (with the Author's sanction) by E. ATKINSON, Ph.D., F.C.S. Seventh Edition, carefully revised; with 37 pages of New Matter, 7 Plates, 569 Woodcuts, and an Appendix of Questions. Crown 8vo, 7*s.* 6*d.*

GIBSON. A TEXT-BOOK OF ELEMENTARY BIOLOGY. By R. J. HARVEY GIBSON, M.A., F.R.S.E., Lecturer on Botany in University College, Liverpool. With 192 Illustrations. Crown 8vo, 6*s.*

GOODEVE.—*WORKS by T. M. GOODEVE, M.A., Barrister-at-Law; Professor of Mechanics at the Normal School of Science and the Royal School of Mines.*

PRINCIPLES OF MECHANICS. New Edition, re-written and enlarged. With 253 Woodcuts and numerous Examples. Crown 8vo, 6*s.*

THE ELEMENTS OF MECHANISM. New Edition, re-written and enlarged. With 342 Woodcuts. Crown 8vo, 6*s.*

A MANUAL OF MECHANICS: an Elementary Text-Book for Students of Applied Mechanics. With 138 Illustrations and Diagrams, and 141 Examples taken from the Science Department Examination Papers, with Answers. Fcp. 8vo, 2*s.* 6*d.*

HELMHOLTZ.—*WORKS by HERMANN L. F. HELMHOLTZ, M.D., Professor of Physics in the University of Berlin.*

ON THE SENSATIONS OF TONE AS A PHYSIOLOGICAL BASIS FOR THE THEORY OF MUSIC. Second English Edition; with numerous additional Notes, and a new Additional Appendix, bringing down information to 1885, and specially adapted to the use of Musical Students. By ALEXANDER J. ELLIS, B.A., F.R.S., F.S.A., &c., formerly Scholar of Trinity College, Cambridge. With 68 Figures engraved on Wood, and 42 Passages in Musical Notes. Royal 8vo, 28*s.*

POPULAR LECTURES ON SCIENTIFIC SUBJECTS. With 68 Woodcuts. 2 Vols. crown 8vo, 15*s.*, or separately, 7*s.* 6*d.* each.

HERSCHEL. OUTLINES OF ASTRONOMY. By Sir JOHN F. W. HERSCHEL, Bart., K.H., &c., Member of the Institute of France. Twelfth Edition, with 9 Plates, and numerous Diagrams. 8vo, 12*s.*

HUDSON AND GOSSE. THE ROTIFERA OR 'WHEEL ANIMALCULES.' By C. T. HUDSON, LL.D., and P. H. GOSSE, F.R.S. With 30 Coloured and 4 Uncoloured Plates. In 6 Parts. 4to, price 10*s.* 6*d.* each; Supplement, 12*s.* 6*d.* Complete in Two Volumes, with Supplement, 4to, £4 4*s.*

** The Plates in the Supplement contain figures of almost all the Foreign Species, as well as of the British Species, that have been discovered since the original publication of Vols. I. and II.

IRVING. PHYSICAL AND CHEMICAL STUDIES IN ROCK-METAMORPHISM, based on the Thesis written for the D.Sc. Degree in the University of London, 1888. By the Rev. A. IRVING, D.Sc. Lond., Senior Science Master at Wellington College. 8vo, 5*s.*

JORDAN. THE OCEAN: A Treatise on Ocean Currents and Tides and their Causes. By WILLIAM LEIGHTON JORDAN, F.R.G.S. 8vo, 21*s.*

KOLBE. A SHORT TEXT-BOOK OF INORGANIC CHE-MISTRY. By Dr. HERMANN KOLBE, late Professor of Chemistry in the University of Leipzig. Translated and Edited by T. S. HUM-PIDGE, Ph.D., B.Sc. (Lond.), late Professor of Chemistry and Physics in the University College of Wales, Aberystwyth. New Edition. Revised by H. LLOYD-SNAPE, Ph.D., D.Sc. (Lond.), Professor of Chemistry in the University College of Wales, Aberystwyth. With a Coloured Table of Spectra and 66 Woodcuts. Crown 8vo, 8*s.* 6*d.*

LADD.—*WORKS by GEORGE T. LADD, Professor of Philosophy in Yale University.*

ELEMENTS OF PHYSIOLOGICAL PSYCHOLOGY: A TREATISE OF THE ACTIVITIES AND NATURE OF THE MIND FROM THE PHYSICAL AND EXPERI-MENTAL POINT OF VIEW. With 113 Illustrations. 8vo, price 21*s.*

OUTLINES OF PHYSIOLOGICAL PSYCHOLOGY. With numerous Illustrations. 8vo. 12*s.*

LARDEN. ELECTRICITY FOR PUBLIC SCHOOLS AND COLLEGES. With numerous Questions and Examples with Answers, and 214 Illustrations and Diagrams. By W. LARDEN, M.A. Crown 8vo, 6*s.*

LINDLEY AND MOORE. THE TREASURY OF BOTANY, OR POPULAR DICTIONARY OF THE VEGETABLE KINGDOM : with which is incorporated a Glossary of Botanical Terms. Edited by J. LINDLEY, M.D., F.R.S., and T. MOORE, F.L.S. With 20 Steel Plates, and numerous Woodcuts. 2 Parts, fcp. 8vo, price 12s.

LOUDON. AN ENCYCLOPÆDIA OF PLANTS. By J. C. LOUDON. Comprising the Specific Character, Description, Culture, History, Application in the Arts, and every other desirable particular respecting all the plants indigenous to, cultivated in, or introduced into, Britain. Corrected by Mrs. LOUDON. 8vo, with above 12,000 Woodcuts, price 42s.

MARTIN. NAVIGATION AND NAUTICAL ASTRONOMY. Compiled by Staff-Commander W. R. MARTIN, R.N., Instructor in Surveying, Navigation, and Compass Adjustment ; Lecturer on Meteorology at the Royal Naval College, Greenwich. Sanctioned for use in the Royal Navy by the Lords Commissioners of the Admiralty. Royal 8vo, 18s.

MENDELÉEFF. THE PRINCIPLES OF CHEMISTRY. By D. MENDELÉEFF, Professor of Chemistry in the University of St. Petersburg. Translated by GEORGE KAMENSKY, A.R.S.M. of the Imperial Mint, St. Petersburg, and Edited by A. J. GREENAWAY, F.I.C., Sub Editor of the Journal of the Chemical Society. With 97 Illustrations. 2 Vols. 8vo, 36s.

MEYER. OUTLINES OF THEORETICAL CHEMISTRY. By LOTHAR MEYER, Professor of Chemistry in the University of Tübingen. Translated by Professors P. PHILLIPS BEDSON, D.Sc., and W. CARLETON WILLIAMS, B.Sc. 8vo, 9s.

MILLER.—*WORKS by WILLIAM ALLEN MILLER, M.D., D.C.L., LL.D., late Professor of Chemistry in King's College, London.*

THE ELEMENTS OF CHEMISTRY, Theoretical and Practical.

PART II. INORGANIC CHEMISTRY. Sixth Edition, revised throughout, with Additions by C. E. GROVES, Fellow of the Chemical Societies of London, Paris, and Berlin. With 376 Woodcuts. 8vo, price 24s.

PART III. ORGANIC CHEMISTRY, or the Chemistry of Carbon Compounds. *Hydrocarbons, Alcohols, Ethers, Aldehides and Paraffinoid Acids.* Fifth Edition, revised and in great part re-written, by H. E. ARMSTRONG, F.R.S., and C. E. GROVES, F.C.S. 8vo, price 31s. 6d.

MITCHELL. MANUAL OF PRACTICAL ASSAYING. By JOHN MITCHELL, F.C.S. Sixth Edition. Edited by W. CROOKES, F.R.S. With 201 Woodcuts. 8vo, 31*s*. 6*d*.

MORGAN. ANIMAL BIOLOGY. An Elementary Text Book. By C. LLOYD MORGAN, Professor of Animal Biology and Geology in University College, Bristol. With numerous Illustrations. Crown 8vo, 8*s*. 6*d*.

ODLING. A COURSE OF PRACTICAL CHEMISTRY, Arranged for the use of Medical Students, with express reference to the Three Months' Summer Practice. By WILLIAM ODLING, M.A., F.R.S. With 71 Woodcuts. Crown 8vo, 6*s*.

OLIVER. ASTRONOMY FOR AMATEURS: A PRACTICAL MANUAL OF TELESCOPIC RESEARCH IN ALL LATITUDES ADAPTED TO THE POWERS OF MODERATE INSTRUMENTS. Edited by JOHN A. WESTWOOD OLIVER, with the assistance of T. W. BACKHOUSE, F.R.A.S.; S. W. BURNHAM, M.A., F.R.A.S.; J. RAND CAPRON, F.R.A.S.; W. F. DENNING, F.R.A.S.; T. GWYN ELGER, F.R.A.S.; W. S. FRANKS, F.R.A.S.; J. E. GORE, M.R.I.A., F.R.A.S.; SIR HOWARD GRUBB, F.R.S., F.R.A.S.; E. W. MAUNDER, F.R.A.S.; and others. Illustrated. Crown 8vo, 7*s*. 6*d*.

OSTWALD. SOLUTIONS. By W. OSTWALD, Professor of Chemistry in the University of Leipzig. Being the Fourth Book, with some additions, of the Second Edition of Ostwald's "Lehrbuch der Allgemeinen Chemie." Translated by M. M. PATTISON MUIR, Professor of Gonville and Caius College, Cambridge. 8vo, 10*s*. 6*d*.

PAYEN. INDUSTRIAL CHEMISTRY; A Manual for use in Technical Colleges or Schools, also for Manufacturers and others, based on a Translation of Stohmann and Engler's German Edition of PAYEN'S *Précis de Chimie Industrielle.* Edited and supplemented with Chapters on the Chemistry of the Metals, &c., by B. H. PAUL, Ph.D. With 698 Woodcuts. 8vo, 42*s*.

REYNOLDS. EXPERIMENTAL CHEMISTRY for Junior Students. By J. EMERSON REYNOLDS, M.D., F.R.S., Professor of Chemistry, Univ. of Dublin. Fcp. 8vo, with numerous Woodcuts.

> PART I.—*Introductory,* 1*s*. 6*d*.
>
> PART II.—*Non-Metals,* with an Appendix on Systematic Testing for Acids, 2*s*. 6*d*.
>
> PART III.—*Metals and Allied Bodies,* 3*s*. 6*d*.
>
> PART IV.—*Chemistry of Carbon Compounds,* 4*s*.

PROCTOR.—*WORKS by RICHARD A. PROCTOR.*

OLD AND NEW ASTRO-NOMY. In 12 Parts. Price 2*s.* 6*d.* each; supplementary section, 1*s.*; complete, 36*s.* cloth.

LIGHT SCIENCE FOR LEISURE HOURS; Familiar Essays on Scientific Subjects, Natural Phenomena, &c. 3 Vols. crown 8vo, 5*s.* each.

THE ORBS AROUND US; a Series of Essays on the Moon and Planets, Meteors, and Comets. With Chart and Diagrams, crown 8vo, 5*s.*

OTHER WORLDS THAN OURS; The Plurality of Worlds Studied under the Light of Recent Scientific Researches. With 14 Illustrations, crown 8vo, 5*s.* Cheap Edition, crown, 3*s.* 6*d.*

THE MOON; her Motions, Aspects, Scenery, and Physical Condition. With Plates, Charts, Woodcuts, and Lunar Photographs, crown 8vo, 5*s.*

UNIVERSE OF STARS; Presenting Researches into and New Views respecting the Constitution of the Heavens. With 22 Charts and 22 Diagrams, 8vo. 10*s.* 6*d.*

LARGER STAR ATLAS for the Library. in 12 Circular Maps, with Introduction and 2 Index Pages. Folio, 15*s.*, or Maps only, 12*s.* 6*d.*

NEW STAR ATLAS for the Library, the School, and the Observatory, in 12 Circular Maps (with 2 Index Plates). Crown 8vo, 5*s.*

THE STUDENT'S ATLAS. In 12 Circular Maps on a Uniform Projection and 1 Scale, with 2 Index Maps. Intended as a *vade-mecum* for the Student of History, Travel, Geography, Geology, and Political Economy. With a letter-press Introduction illustrated by several cuts. 5*s.*

ELEMENTARY PHYSICAL GEOGRAPHY. With 33 Maps and Woodcuts. Fcp. 8vo, 1*s.* 6*d.*

LESSONS IN ELEMENTARY ASTRONOMY; with Hints for Young Telescopists. With 47 Woodcuts. Fcp. 8vo, 1*s.* 6*d.*

FIRST STEPS IN GEOMETRY: a Series of Hints for the Solution of Geometrical Problems; with Notes on Euclid, useful Working Propositions, and many Examples. Fcp. 8vo, 3*s.* 6*d.*

EASY LESSONS IN THE DIFFERENTIAL CALCULUS: indicating from the Outset the Utility of the Processes called Differentiation and Integration. Fcp. 8vo, 2*s.* 6*d.*

THE STARS IN THEIR SEASONS. An Easy Guide to a Knowledge of the Star Groups, in 12 Large Maps. Imperial 8vo, 5*s.*

STAR PRIMER. Showing the Starry Sky Week by Week, in 24 Hourly Maps. Crown 4to. 2*s.* 6*d.*

THE SEASONS PICTURED IN 48 SUN VIEWS OF THE EARTH, and 24 Zodiacal Maps, &c. Demy 4to, 5*s.*

ROUGH WAYS MADE SMOOTH. Familiar Essays on Scientific Subjects. Crown 8vo, 5*s.* Cheap Edition, crown, 3*s.* 6*d.*

HOW TO PLAY WHIST: WITH THE LAWS AND ETIQUETTE OF WHIST. Crown 8vo, 3*s.* 6*d.*

HOME WHIST: an Easy Guide to Correct Play. 16mo, 1*s.*

OUR PLACE AMONG INFINITIES. A Series of Essays contrasting our Little Abode in Space and Time with the Infinities around us. Crown 8vo, 5*s.*

[Continued.

PROCTOR.—*WORKS by RICHARD A. PROCTOR—continued.*

STRENGTH AND HAPPI-
NESS. Crown 8vo, 5s.

STRENGTH : How to get Strong
and keep Strong, with Chapters
on Rowing and Swimming, Fat,
Age, and the Waist. With 9 Illustrations. Crown 8vo, 2s.

THE EXPANSE OF HEAVEN.
Essays on the Wonders of the
Firmament. Crown 8vo, 5s.

THE GREAT PYRAMID, OB-
SERVATORY, TOMB, AND
TEMPLE. With Illustrations.
Crown 8vo, 5s.

PLEASANT WAYS IN
SCIENCE. Crown 8vo, 5s.
Cheap Edition, crown, 3s. 6d.

MYTHS AND MARVELS OF
ASTRONOMY. Crown 8vo, 5s.

CHANCE AND LUCK ; a Dis-
cussion of the Laws of Luck,
Coincidences, Wagers, Lotteries,
and the Fallacies of Gambling,
&c. Crown 8vo, 2s. boards,
2s. 6d. cloth.

NATURE STUDIES. By
GRANT ALLEN, A. WILSON,
T. FOSTER, E. CLODD, and
R. A. PROCTOR. Crown 8vo, 5s.

LEISURE READINGS. By E.
CLODD, A. WILSON, T. FOSTER,
A. C. RUNYARD, and R. A.
PROCTOR. Crown 8vo, 5s.

SCOTT. WEATHER CHARTS AND STORM WARNINGS.
By ROBERT H. SCOTT, M.A., F.R.S. With numerous Illustrations.
Crown 8vo, 6s.

SLINGO AND BROOKER. ELECTRICAL ENGINEERING
FOR ELECTRIC LIGHT ARTISANS AND STUDENTS.
(Embracing those branches prescribed in the Syllabus issued by the City and
Guilds Technical Institute.) By W. SLINGO, Principal of the Telegraphists'
School of Science, &c., &c., and A. BROOKER, Instructor on Electrical
Engineering at the Telegraphists' School of Science. With 307 Illustrations.
Crown 8vo, 10s. 6d.

SMITH. GRAPHICS ; OR, THE ART OF CALCULATION
BY DRAWING LINES, applied to Mathematics, Theoretical Me-
chanics and Engineering, including the Kinetics and Dynamics of Machinery,
and the Statics of Machines, Bridges, Roofs, and other Engineering Structures.
By ROBERT H. SMITH, Professor of Civil and Mechanical Engineering,
Mason Science College, Birmingham.

PART I. Text, with separate Atlas of Plates—Arithmetic, Algebra,
Trigonometry, Vector, and Locor Addition, Machine Kinematics, and Statics
of Flat and Solid Structures. 8vo, 15s.

THORPE. A DICTIONARY OF APPLIED CHEMISTRY.
By T. E. THORPE, B.Sc. (Vict.), Ph.D., F.R.S., Treas. C.S., Professor of
Chemistry in the Royal College of Science, London. Assisted by Eminent
Contributors. To be published in 3 vols. 8vo. Vols. I. and II. £2 2s. each.
[*Now ready.*

TYNDALL.—*WORKS by JOHN TYNDALL, F.R.S., &c.*

FRAGMENTS OF SCIENCE. 2 Vols. Crown 8vo, 16*s.*

NEW FRAGMENTS. Crown 8vo, 10*s.* 6*d.*

HEAT A MODE OF MOTION. Crown 8vo, 12*s.*

SOUND. With 204 Woodcuts. Crown 8vo, 10*s.* 6*d.*

RESEARCHES ON DIAMAGNETISM AND MAGNE-CRYS-TALLIC ACTION, including the question of Diamagnetic Polarity. Crown 8vo, 12*s.*

ESSAYS ON THE FLOATING-MATTER OF THE AIR in relation to Putrefaction and Infection. With 24 Woodcuts. Crown 8vo, 7*s.* 6*d.*

LECTURES ON LIGHT, delivered in America in 1872 and 1873. With 57 Diagrams. Crown 8vo, 5*s.*

LESSONS IN ELECTRICITY AT THE ROYAL INSTITU-TION, 1875-76. With 58 Woodcuts. Crown 8vo, 2*s.* 6*d.*

NOTES OF A COURSE OF SEVEN LECTURES ON ELECTRICAL PHENOMENA AND THEORIES, delivered at the Royal Institution. Crown 8vo, 1*s.* sewed, 1*s.* 6*d.* cloth.

NOTES OF A COURSE OF NINE LECTURES ON LIGHT, delivered at the Royal Institution. Crown 8vo, 1*s.* sewed, 1*s.* 6*d.* cloth.

WATTS' DICTIONARY OF CHEMISTRY. Revised and entirely Re-written by H. FORSTER MORLEY, M.A.. D.Sc., Fellow of, and lately Assistant-Professor of Chemistry in, University College, London ; and M. M. PATTISON MUIR, M.A., F.R.S.E., Fellow, and Prælector in Chemis y, of Gonville and Caius College, Cambridge. Assisted by Eminent Contributors. To be Published in 4 Vols. 8vo. Vols. I. II. 42*s.* each. Vol. III. 50*s.*

[Now ready.

WEBB. CELESTIAL OBJECTS FOR COMMON TELESCOPES. By the Rev. T. W. WEBB, M.A. Fourth Edition, adapted to the Present State of Sidereal Science ; Map, Plate, Woodcuts. Crown 8vo, price 9*s.*

WILLIAMS. MANUAL OF TELEGRAPHY. By W. WILLIAMS, Superintending Indian Government Telegraphs. With 93 Woodcuts. 8vo, 10*s.* 6*d.*

WRIGHT. OPTICAL PROJECTION : A Treatise on the Use of the Lantern in Exhibition and Scientific Demonstration. By LEWIS WRIGHT, Author of "Light : a Course of Experimental Optics." With 232 Illustrations. Crown 8vo, 6*s.*

5000.7.93.   BRADBURY, AGNEW, & CO. LD., PRINTERS, WHITEFRIARS.

www.ingramcontent.com/pod-product-compliance
Lightning Source LLC
Chambersburg PA
CBHW021524210326
41599CB00012B/1372